X-Ray Lasers

X-RAY LASERS

Raymond C. Elton
NAVAL RESEARCH LABORATORY
WASHINGTON, D.C.

ACADEMIC PRESS, INC.

Harcourt Brace Jovanovich, Publishers

Boston   San Diego   New York
Berkeley   London   Sydney
Tokyo   Toronto

This book is printed on acid-free paper.

ACADEMIC PRESS, INC.
1250 Sixth Avenue, San Diego, CA 92101

United Kingdom Edition published by
ACADEMIC PRESS LIMITED
24–28 Oval Road, London NW1 7DX

The outline of a laser-driven x-ray laser on the cover is adapted from Fig. 20b and Ref. 79 of Chapter 2 with kind permission of the authors and journal.

The material contained in this book does not represent the policies or views of the U.S. Naval Research Laboratory or the Department of the Navy.

Library of Congress Cataloging-in-Publication Data

Elton, Raymond C., Date–
    X-ray lasers/Raymond C. Elton.
        p. cm.
    Includes bibliographies and index.
    ISBN 0-12-238080-0 (alk. paper)
    1. X-ray lasers. I. Title.
TA1707.E37   1990
621.3′6--dc20

PRINTED IN THE UNITED STATES OF AMERICA
90 91 92 93   9 8 7 6 5 4 3 2 1

To Patty,
  Allison, Audrey and Adrienne

Contents

Chapter 3 Pumping by Exciting Plasma Ions 99

Chapter 6   Alternate Approaches   199

Chapter 7 Summary, Applications, and Prognosis

Preface

Extensive advances have been made in extending lasers towards the x-ray region since the invention of the device. This is particularly true experimentally over the last 5 years. However, there is at this point no single reference book that provides a broad survey of the fundamentals in x-ray lasers, yet contains sufficient detail to be of interest to the active researcher. This volume is an attempt to fill this void. It is intended to serve both as a primer for the reader entering the field and as a convenient reference source for the laser scientist familiar with the field. It is not intended to be a review of activity in this field, complete with all references to previous and ongoing research. For such a general literature overview, the reader is referred particularly to the comprehensive 1976 article[1] and the 1984 citation summary[2] listed at the end of the Preface. Other reviews on specific aspects of the field are cited at appropriate points in the text.

The main emphasis in this book is on x-ray lasing created using high temperature plasmas as the medium. This is the area in which the most success has occurred to date. Research on other possible media also are presently pursued to a lesser extent. Some efforts are active and some are merely conjecture at this stage. These are discussed in Chapter 6.

Throughout this book wavelengths are given in Angstrom units (Å) when longer than 12.4 Å (12.4×10^{-8} cm). Besides being traditional for the x-ray spectral region and used extensively in the x-ray laser literature, this results in conveniently sized quantities. Photon energy (hν) is used for shorter wavelengths, i.e., energies above 1 keV. The approximate conversion here is λ [Å] = 12400/hν [eV]. Micrometers (μm) are used in the infrared spectral region to describe driving lasers. For other quantities, cgs units are used.

In the first chapter, some fundamental definitions and practices important to x-ray laser research are reviewed. In Chapter 2, the basic principles of designing, operating, and diagnosing x-ray lasers are described. Chapters

3–5 emphasize particular pumping methods for plasma x-ray lasers. Analysis of each promising scheme is followed by a description of experiments, where they exist. As mentioned above, Chapter 6 describes several alternative (to plasma) approaches currently under investigation. In Chapter 7 the material is summarized and x-ray lasers are compared with other high brightness sources. Numerous potential applications are also described in this final chapter. A list of symbols and the locations where each is first defined follows Chapter 7.

The author wishes to thank his many colleagues at the Naval Research Laboratory; the Lawrence Livermore National Laboratory; the National Institute of Standards and Technology and the University of Maryland; and others elsewhere, whose collaborative contributions are recalled with appreciation. The author takes particular pleasure in expressing his appreciation to Professor H. R. Griem, who provided him with excellent training in the field of plasma spectroscopy as well as continued inspiration and collaboration in the x-ray laser specialty. Very special thanks go to the author's wife and family for their patience, encouragement and understanding while this book was in preparation.

REFERENCES

1. R. W. Waynant and R. C. Elton, *Proc. IEEE* **64**, 1059 (1976).
2. P. L. Hagelstein, in "Atomic Physics 9," R. S. Van Dyck, Jr., and E. N. Fortson, eds., p. 382 (World Scientific Publ. Co., Ltd., Singapore, 1984).

Chapter 1 | Introduction

1. A LASER FOR THE X-RAY REGION

1.1. Background

Beginning with the initial papers on lasers in the early 1960s, there has been anticipation of the eventual extension of lasing into the x-ray region. The authors recognized that the challenges facing such an extension would be formidable. This was basically because of what were perceived to be: (a) a generally unfavorable scaling of the gain with wavelength, (b) an inefficiency of cavity reflectors, and (c) the large pumping power that would be required. There were also early concerns that lasing would be necessarily self terminating on a femtosecond (10^{-15} seconds) time scale characteristic of traditional Kα-transition decay rates. At that time such short pulse intervals seemed virtually impossible to achieve.

However, it was subsequently realized that quasi-continuous-wave (cw) operation, limited only by the duration of the pumping impulse, could be achieved using outershell "optical" transitions in highly ionized atoms. In this mode, the population inversion is maintained by rapid depletion of the lower-state density. This is usually accomplished through radiative decay to the ground state. This concept developed as an extrapolation from successful visible and near-ultraviolet lasers using atoms in low stages of ionization. This is the area in which the most progress has taken place over the past 20 years. There remains hope that eventually progression to innershell transitions will lead to even shorter wavelengths in lower ionization stages, and perhaps even in neutral atoms.

1

1.2. The Data Base

Fortunately, a vast amount of information on the atomic structure and radiation characteristics of highly charged ions already existed from very early astrophysical and plasma studies. This was augmented by laboratory simulation experiments, beginning in the 1930s with vacuum sparks. Experimental work on magnetically-confined controlled-thermonuclear fusion, and in particular the study of impurity characteristics and diagnostics, contributed immensely to the data and expertise bases. However, it was the developments in the inertial-confinement fusion programs, beginning in the early 1970s, that provided the theory, data and eventually the high volumetric power densities needed for successful x-ray lasing experiments.

Such highly stripped ions that mimic neutral atoms are considered to belong to a particular isoelectronic sequence and have certain common characteristics. The most easily extrapolated of these is the hydrogenic sequence of atoms stripped to one remaining electron. The extrapolation formulae are particularly simple and accurate. Hence this has been a favorite sequence in x-ray laser design, and will be used extensively for illustration in this book. Further sequences involving two, three, or more remaining bound electrons are designated helium-like (He-like), lithium-like (Li-like), etc.

A particular ion is usually designated by the number of electrons that have been removed, i.e., C^{5+} for a hydrogenic carbon ion of nuclear charge $Z = 6$. We will use this convention throughout most of this book. Spectra are often labeled by roman numerals, beginning with I for the neutral atom. This convention is particularly prominent in the astrophysical literature. Hence, spectra of hydrogenic carbon would be designated by C VI. This becomes somewhat clumsy for highly charged heavy atoms. Where it is necessary to use this notation, we will relate the two to avoid confusion.

1.3. The Lasing Medium

Ions, along with their stripped (free) electrons in a localized region (overall neutral), are said to form a plasma. This is sometimes referred to as the fourth state of matter. The plasma is mainly defined by the temperatures and densities for the electrons and ions, as well as by the size and shape. These parameters are commonly designated by T_e, T_i, N_e and N_i, respectively. Often, but not always, $T_e = T_i$. For a pure single-species plasma, N_e and N_i are related by the charge of the ion. For example, for hydrogenic ions of nuclear charge Z, $N_e = (Z - 1)N_i$. We will use such a relation extensively in the analyses in Chapters 3–5.

1.4. Configurations

Most lasers operate in the visible and infrared spectral regions along an axis defined by a Fabry-Perot style of reflecting cavity. The lasant medium is pumped to a population-density inversion. Amplification along this axis is provided by stimulated emission. The intensity exponentiates further with each pass through the medium. A high degree of collimation and coherence is achieved with efficient cavities.

In the x-ray region, the reflectivity of available cavity mirrors is relatively poor, as detailed below. Hence, x-ray lasers to date have been based mainly on amplification of inherent spontaneous emission (dubbed ASE) throughout the medium. This occurs in a single pass along a particular direction. (A more complete definition of ASE is presented in Section 2.2 of Chapter 2.) This requires gain coefficients in the medium approximately 100 times larger than those for more familiar lasers operating with efficient resonators. It is intriguing to consider that a properly prepared medium can, in principle, undergo laser action in any direction. This is the basis for suggestions that certain astronomical objects may be radiating by ASE.

However, if significant enhancement is to be expected, a preferred direction must be defined. This could be accomplished in a spherical plasma, for example, by an externally applied driver beam from a low power oscillator. This is sometimes referred to as the MOPA (Master-Oscillator, Power-Amplifier) concept. Usually, however, other constraints dictate that the lasing geometry be rod-shaped. These include the available driving power as well as reabsorption associated with opacity on the lower-level depletion transition. Such constraints will be discussed at length in Chapter 2.

1.5. Phases of Development

The logical progression of an x-ray laser from concept to application is depicted in the flowchart in Fig. 1. Numerical modeling closely parallels the experimental efforts. Such stages will be recognized throughout the book.

We may illustrate the flowchart definitions with some rather general examples. The extensive conceptualization which occurred during the 1970s is an example of concentration in the top half. It has been in the 1980s that advancements into gain experiments and devices have been made, as appropriate plasma media became available. Harmonic generation starting with long wavelength lasers has reached the feasibility-experiment stage, as far as producing useful coherent emission at short wavelengths. On the other hand, free-electron x-ray and nuclear gamma-ray lasers are presently in the concept

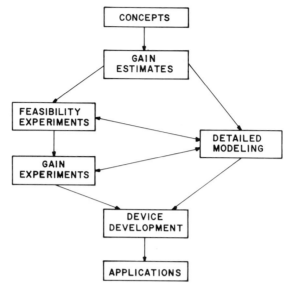

Fig. 1. Logical progression of x-ray laser research and development.

stage at present. Other alternatives such as solid-state crystal x-ray lasing, stimulated electron capture and Compton scattering also are conceptual at present.

2. DEFINITIONS AND OPTICAL PROPERTIES UNIQUE TO THE REGION OF INTEREST

Before proceeding to the main body of this volume, it is important to understand certain definitions and characteristics unique to the short-wavelength spectral region of interest. These definitions are essential to the discussion of x-ray lasers. We begin by defining the wavelength region over which the principles apply. Following this are discussed certain unique optical properties of matter for the short wavelength spectral region. These concepts are very important for understanding the added complexities associated with lasers in that spectral class. Instrumentation for spectroscopic measurement, such as gratings and detectors, are discussed in Section 6 of Chapter 2.

2.1. Wavelength Ranges Defined

Beginning with some earlier recommendations proposed by Samson[1] we can categorize ultraviolet wavelengths shorter than 2000 Å (1 Å or Angstrom =

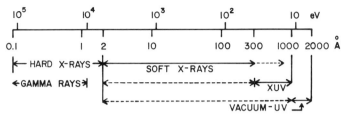

Fig. 2. Wavelength and photon energy ranges and designations useful for x-ray lasers.

10^{-8} cm) into various somewhat-overlapping subdivisions. The boundaries are not at all rigid, but are those traditionally accepted in the x-ray laser, plasma, and astrophysical fields. These are illustrated in Fig. 2.

2.1.1. The Vacuum-UV Region

The broad range extending to 2 Å, in which the atmosphere does not transmit over a useful path length, has been named the vacuum ultraviolet or "vacuum-uv" or even "vuv" spectral region. Propagation over useful distances is only possible in vacuum, or, for the longer wavelength end, in the lighter rare gases. With the further labeling of the region short of 1000 Å (see below), "vacuum-uv" has come to represent particularly the 1000–2000 Å range, where traditional optics are available. This explains the use of a combined dashed/solid vacuum-uv line in Fig. 2.

2.1.2. The Xuv (or Euv) Region

At wavelengths shorter than 1000 Å, and particularly for the 300–1000 Å vacuum range, the wavelength region is named the extreme ultraviolet or "xuv" (or euv). The designation xuv seems to be preferred over euv, perhaps because of an association with the adjacent soft x-ray region.

2.1.3. The X-Ray Region

The total "x-ray" region broadly covers the 700 Å down to 0.1 Å wavelength range[2,3]. However, it is traditionally considered to begin at about 300 Å, where the xuv range ends. This x-ray region is further categorized as "soft" for wavelengths in the ~ 2–300 Å range and "hard" for shorter wavelengths. X-rays in the soft range are typically described by wavelengths in Angstrom units. This custom will be used throughout most of this book, because x-ray lasers are currently operated in this wavelength range. Wavelengths convert to frequency through $\lambda = c/v$, where c is the velocity of light in vacuum. For hard x-rays it is traditional and somewhat more practical to use the photon energy hv in keV units as a label. The conversion is

$$hv = \frac{12.4}{\lambda} \quad \text{keV}, \tag{1}$$

for a wavelength λ in Angstrom units. Hence, the $\lambda = 0.1$ to 2 Å hard x-ray region is equivalent to $hv = 6.2$–120 keV. This is indicated by the upper scale in Fig. 2.

2.1.4. The Hard X-Ray and Gamma-Ray Regions

Photons emitted from radioactive nuclei with energies hv in the broad 10 keV to 250 MeV range are called gamma-rays[3]. The definition overlaps the "hard" x-ray range at the lower energies. The distinction is between nuclear and atomic transitions, particularly for the material covered in this book. Hence, currently conceived 10–100 keV nuclear gamma-ray lasers described in Chapter 6 could also quite properly be classified as x-ray lasers.

2.2. Windows and Filters

In the vacuum-uv spectral region, as defined above, windows are necessary when the radiation source operates at such a high pressure that available vacuum pumps cannot reduce the pressure in the detector system sufficiently for radiation to be transmitted to the detector. This depends on the particular gas and the wavelength[1]. For example, an inert gas such as helium has good transmission at a reasonably high pressure for wavelengths as short as the photoionization edge at 539 Å. As such, it can often be used in a purge mode at atmospheric pressure over a moderate path length (molecular helium absorption can occur at high pressures for path lengths $\gtrsim 1$ m). On the other hand, diatomic gases such as nitrogen and oxygen (air) are highly opaque for wavelengths shorter than 2000 Å and must be removed. The window therefore isolates the detection system from the source. Gases can also serve to filter out undesirable radiation at wavelengths shorter than the transmission cutoff. For example, argon at a pressure of ~ 1 Torr attenuates wavelengths shorter than ~ 800 Å. A somewhat less-attractive alternative to a window is an open pinhole or a narrow slit, described in Section 2.2.3 below.

2.2.1. Crystalline

A number of dielectric materials remain transparent in the vacuum-uv spectral region. They can be used for windows, lenses, and $\frac{1}{4}$ and $\frac{1}{2}$ wave plates (if the material is birefringent). The transmission remains high near (within ~ 50 Å) the cutoff wavelength. The wavelength dependence of the transmission of various dielectric materials[4] is shown in Fig. 3. As popular examples, high quality quartz can be used to wavelengths as short as 1650 Å. In fact, 80% transmission has been reported at 1500 Å with cultured crystals[5]. Crystalline lithium-fluoride transmits to a wavelength as short as 1100 Å. (However, absorption increases and the cutoff effectively shifts to longer wavelengths if color centers form because of plasma bombardment.) Again, such short-wavelength cutoffs in transmission are not altogether detrimental when they serve to eliminate higher-order spectral interference in analyzing many-line data.

Such windows can also be used as substrates to support multilayer bandpass filters for the regions in which they transmit. Aluminum, lithium fluoride and magnesium fluoride have been successfully used[6] as such substrates. Peak transmissions are 15–30%, with half-widths of 150–400 Å.

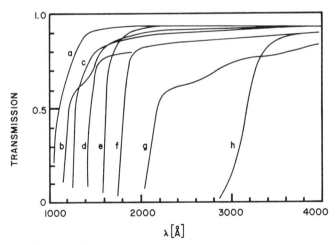

Fig. 3. Transmittances of various ultraviolet optical materials: (a) LiF, 2 mm; (b) MgF$_2$; (c) CaF$_2$, 1 mm; (d) sapphire, 1 mm; (e) quartz, 1 mm; (f) ADP, 2.5 mm; (g) calcite, 11 mm; (h) microscope slide glass, 1.3 mm. (From Ref. 4.)

2.2.2. Thin Metal

At shorter xuv wavelengths some materials, known as free-electron metals, behave similarly to a dense plasma and are useful as broadband filters in the form of unbacked thin films. The transmission band extends from a short wavelength absorption-edge cutoff to the plasma cutoff at long wavelengths, above which reflection dominates. In between, the photon penetration depth is large. If the sample is sufficiently thin (~ 1000 Å), transmittances as high as 70% are possible. Such transmittances[6] for various materials and thicknesses are plotted in Fig. 4. Such films are extemely delicate. However, they have been used successfully over narrow slit- or pinhole-openings and with fine-mesh screen backings at pressure differences as high as an atmosphere. They are not as convenient as thick windows, particularly for large aperture applications.

A particularly popular example is aluminum, which transmits between 170 Å and ~ 700 Å and is quite readily available in the form of thin films. Beryllium and carbon are other favorites (not included in Fig. 4) for wavelengths longer than the K-absorption edges at 111 Å and 43.6 Å, respectively.

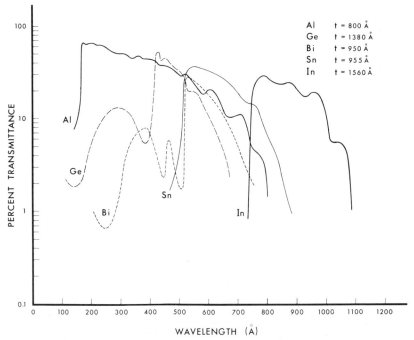

Fig. 4. Transmittances of some thin metal films of thicknesses shown. (From Ref. 6.)

Beryllium (and boron) has the advantage that it can be used in thicker and more rugged layers because of its low mass. However, beryllium must be handled with caution because of toxicity. Just as with crystalline windows, such thin films can serve as bandpass filters, very convenient for rejecting undesirable emission from spectral regions outside that of interest.

2.2.3. Open Structure

An alternative to windows for isolating regions of different pressure is high-speed differential pumping on the low pressure side through a slit or other small aperture. While sometimes cumbersome, this can be as effective as a thin film over a similar opening and is an alternative when no thin-film window/filter is available for the particular wavelength of interest. Pinhole cameras for x-ray diagnostics, discussed in Section 6.1.1 of Chapter 2 are of similar nature.

2.3. Reflectors

The optical properties of reflectors in the vacuum-uv regions are described here. Cavity and spectrographic designs depending on such reflectors are treated later (Sections 2.3 and 5.5 of Chapter 2).

2.3.1. Metallic Coatings

2.3.1a. NORMAL INCIDENCE.   For high reflectance in the vacuum-uv at normal incidence to the surface, overcoated aluminum is the clear choice for the 1000–2000 Å region. Reflectances of 80–90% can be maintained to wavelengths as short as 1200 Å with a fresh layer of aluminum, overcoated with magnesium fluoride to reduce oxidation. This can even be extended to ~ 1000 Å with a lithium-fluoride overcoating[6]. Multilayer dielectric coating technology has been extended to the vacuum-uv region also. Greater than 90% reflectivity can be achieved in the 1450 Å to 2000 Å range. Indeed, it can be as high as 96% and 98% at 1600 Å and 1900 Å, respectively, for use in vacuum-uv cavities[7].

At the shorter xuv wavelengths of primary interest here, all materials are absorbing to various degrees and reflectivity is markedly reduced. Some metals have reflectances in the 20% range[6], as shown in Fig. 5. Platinum coatings are a popular choice over a fairly broad wavelengths range. Other

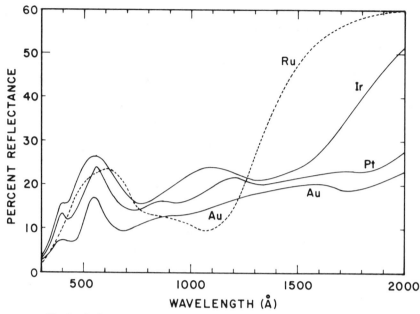

Fig. 5. Reflectance versus wavelength for several materials. (From Ref. 6.)

materials not shown in this figure but of similar reflectance include W, Os, Rh, and Re.

Between 600 Å and 1000 Å, silicon carbide is an extremely hard material of high thermal conductivity with a significantly higher ($\sim 40\%$) reflectance[6]. It is particularly attractive for cavity mirrors, where photon and plasma bombardment could be severe.

Multilayer coatings have been proposed in an effort to extend the wavelength range of high normal-incidence reflectance. Both Pt and Ir, overcoated with aluminum, are two examples of possible stable combinations[6].

2.3.1b. GRAZING INCIDENCE. The above discussion is for normal incidence of the photon beam on the reflecting surface. For wavelengths shorter than 500 Å, useful reflectances have been achieved mostly at a grazing angle of incidence of a few degrees. This is based on total internal reflection at a surface. Reflectances measured at $K\alpha$ wavelengths extending from 114 Å down to 8.3 Å for a gold surface are shown in Fig. 6 as a function of the grazing angle[8].

The main problem with grazing-incidence optics is the severe astigmatism entailed. Such a grazing incidence configuration enters into imaging as well

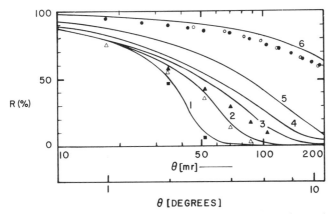

Fig. 6. Reflectivity of a gold surface, measured and calculated, versus grazing angle θ: 1($\blacksquare$) = 8.34 Å; 2($\triangle$) = 13.3 Å; 3($\blacktriangle$) = 23.6 Å; 4 = 44.7 Å; 5 = 64.4 Å; 6($\bigcirc$, $\bullet$) = 114 Å; ———— = theory: (Adapted from Ref. 8.)

as grating spectroscopy (Sections 6.1.3 and 6.2.2 of Chapter 2, respectively). Combinations of mirrors or a mirror and a grating are used to compensate for the astigmatic losses. Grazing incidence optics do not readily adapt to x-ray cavity design (see Section 2.3.1b of Chapter 2).

2.3.2. Layered Synthetic Microstructures

Multilayer reflectors with numerous layers provide improved normal-incidence reflectivity at xuv wavelengths[9–11]. They consist of alternating half-wave-thick layers of highly and slightly absorbing materials, sometimes of equal thickness. The more absorbent metal layers are placed at nodal positions in the standing-wave pattern produced by the superposition of the incident and reflected waves. For an optimized mirror, i.e., with a large reflectivity, two materials are chosen which have the largest possible difference in extinction coefficients.

2.3.2a. MATERIALS. Carbon (and sometimes beryllium and boron) is a popular choice for the spacer layer. This is because carbon has been found to form stable boundaries with practically all heavy metals. Suitable metals are Au, Pt, Ni, Rh, W and a ReW combination. (It is important to choose metals with which carbon does not interdiffuse.) Such coatings are referred to as "layered synthetic microstructures" or LSMs. They are essentially synthetic crystals,

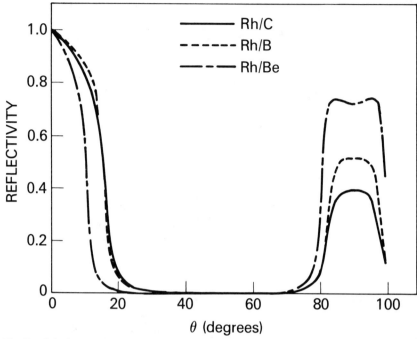

Fig. 7.   Calculated reflectance versus grazing angle for three LSMs at a wavelength of 124 Å. Normal incidence is 90°. (From Ref. 11.)

i.e., they can be thought of as ideal crystals. Indeed, they can be used for x-ray dispersion (see Section 6.3 of Chapter 2). However, carbon has a major absorption edge near 44 Å, which reduces the x-ray reflectance drastically.

2.3.2b. REFLECTIVITIES.   A calculation[11] of reflectivity for three LSMs designed for a wavelength of 124 Å is shown in Fig. 7. The angle plotted along the abscissa here is measured from the mirror surface, i.e., is the grazing angle. This illustrates the high reflectivities predicted near normal incidence. (The increase at grazing incidence is typical of total reflection in any material.) The measured values of reflectivity are considerably less than theoretical predictions. This is particularly true in the xuv spectral region (see Fig. 8 described below). The additional losses can be attributed, at least in part, to the degree of interfacial roughness remaining on consituent layers. The Re/C combination has been found to produce the sharpest boundaries.

Measured peak reflectances for tungsten-carbon (W/C) LSMs are collected in Fig. 8 as a function of photon energy and wavelength[12]. This clearly shows the advantage of shorter wavelengths, for LSM x-ray laser cavities.

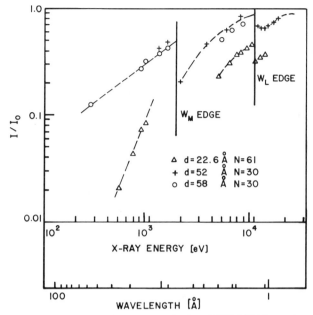

Fig. 8. Measured peak reflectance for a tungsten/carbon (W/C) LSM for various spacings d in Å and number of layers N. (From Ref. 12.)

2.3.2c. APPLICATIONS TO X-RAY LASERS. For cavity operation, LSMs promise a significant improvement in reflectance for the x-ray region. Both angular spread and bandwidth are narrow[13]. Whether or not they can survive short, intense bursts of x-rays and particles from a plasma x-ray laser remains problematic at this time. Besides ruggedness, other areas of research towards improving LSMs include substrate character, process control, and selection of alternate and improved materials.

Such LSM coatings are now being applied to gratings as well as to mirrors[13]. The goal is to extend normal-incidence spectroscopy to the 25–300 Å region, where it is now necessary to use grazing incidence techniques. This would greatly simplify spectroscopic operations in this spectral region and also increase the photon throughput with larger apertures.

REFERENCES

1. J. A. R. Samson, "Techniques of Vacuum Ultraviolet Spectroscopy" (Pied Publications, Inc., Lincoln, Nebraska, 1967).
2. J. A. Bearden, Reviews of Modern Physics **39**, 78 (1967).

3. Handbook of Chemistry and Physics, current edition (CRC Press, Inc., Cleveland, Ohio).
4. J. F. Reintjes, "Coherent Ultraviolet and Vacuum Ultraviolet Sources", in Laser Handbook, M. Bass and M. L. Stitch, eds. (North-Holland, New York, 1985).
5. G. Hass and W. R. Hunter, *Applied Optics* **17,** 2310 (1978).
6. W. R. Hunter, "A Review of Vacuum Ultraviolet Optics", SPIE Proc. No. 140: Optical Coatings II, p. 122 (1978).
7. J. B. Marling and D. B. Lang, *Appl. Phys. Lett.* **31,** 181 (1977).
8. B. L. Henke, in AIP Conference Proceedings No. 75, "Low Energy X-ray Diagnostics", D. T. Atwood and B. L. Henke, eds., p. 85, (American Institute of Physics, New York, 1981).
9. E. Spiller and A. Segmüller, *Annals New York Acad. Sci.* **342,** 188 (1980).
10. Recent status collected in "Multilayer Structures and Laboratory X-Ray Laser Research", N. M. Ceglio and P. Dhez, eds., *SPIE Proceedings* **688,** 76 (1986).
11. T. W. Barbee, Jr., in "X-Ray Microscopy", G. Schmahl and D. Rudolph, eds., p. 144 (Springer-Verlag, New York, 1984).
12. T. W. Barbee, Jr., in AIP Conference Proceedings No. 75, "Low Energy X-ray Diagnostics", D. T. Attwood and B. L. Henke, eds., p. 131 (American Institute of Physics, New York, 1981).
13. W. R. Hunter, in "Technologies for Optoelectronics", R. F. Potter and J. M. Bulabois, eds., *SPIE Proceedings* **869,** 123, (1988).

Chapter 2 | Principles of Short Wavelength Lasers

1. OPERATING MODES

In the soft x-ray spectral region cavity mirrors have poor reflectivity as discussed above, at least by laser standards. Hence, again, it is usually considered necessary to achieve the desired amplification on a single pass of a photon through the medium. This mode of operation is referred to either as "mirrorless" lasing or "amplified spontaneous emission." We will continue with the latter terminology and the associated acronym ASE.

There has been considerable inconsistency in the literature, mostly of a semantic nature[1], between ASE and two other more exotic modes of operation, namely "superradiance" and "superfluorescence." Therefore, before proceeding with a formulation of ASE principles, it is worthwhile to briefly describe these other modes for clarification.

1.1. Dicke Superradiance

In 1954, well before the invention of the laser, R. H. Dicke suggested[2] that a sizable density N_a of atoms confined to a small volume may oscillate with a very high degree of coherence between the individual dipole vibrations. In such a case, the macroscopic dipole moment will be $N_a\mu_a$, where μ_a is the moment of a single atom. The atoms here are strongly coupled by their common radiation fields. The coherent power will then be proportional to $(N_a\mu_a)^2$. This would be distinguishable from spontaneous emission at a rate proportional to N_a. Such a burst of coherent radiation would have a duration proportional to $1/N_a$, rather than an exponential decay with a lifetime independent of N_a. This type of emission, with strong coherence from prepared superposition states and in the absence of de-phasing relaxations, has become known as "Dicke superradiance".

Next suppose that the atoms are prepared incoherently, such that all atomic dipoles are randomly phased and in a partially population-density-inverted state. Dicke, in the same paper, pointed out that the atoms would still be radiatively coupled, providing the volume occupied was $\ll \lambda^3$ for a radiating wavelength λ. Hence, such an inverted system could develop a sizable coherent macroscopic polarization and would radiate again as $(N_a \mu_a)^2$ in a superradiant burst. The dimensions required here are too restrictive to observe this effect, but it does serve to introduce the behavior that has come to be known as "superfluorescence" to be described next.

1.2. Superfluorescence

Superfluorescence is an extension of incoherently prepared Dicke superradiance just described. In this case, however, the atoms are spread over a volume large compared to λ^3. Rather specialized conditions are required, namely, very strong and narrow atomic transitions and very large inversion densities. This is necessary so that the radiative coupling between the atoms remains strong over the larger dimensions. Stated differently, the sample length must not be any larger than the distance radiation travels in one inverse atomic linewidth. In that case, the atoms communicate strongly with each other in a time short compared to a dephasing time. The net observation would then be a burst of intensity again proportional to N_a^2. This emission emerges roughly according to the aspect ratio in a narrow cone from each end of a pencil-like atomic array of Fresnel number $\mathscr{F} \equiv r^2/L\lambda \approx 1$, for a length L and radius r.

1.3. Amplified Spontaneous Emission (ASE)

Amplified spontaneous emission can, in the present context, be understood as subthreshold superfluorescence. The term and associated acronym ASE refer to the situation depicted[3] in Fig. 1. In this mode of operation, the spontaneous emission from a group of inverted atoms is linearly amplified by the same atoms, with an appreciable gain in at least one direction. However, the gain, transition rate, and line width are not adequate for generating the more complex superfluorescent effect described above.

A system can, in principle, change from an ASE mode to superfluorescence and perhaps even superradiance. For example, enhanced line narrowing can be associated with extended lifetimes of the upper-laser level.

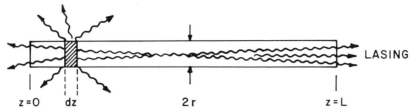

Fig. 1.   Schematic of an amplified-spontaneous-emission or ASE laser. (From Ref. 3.)

Population inversions created in times much shorter than this could produce genuine Dicke superradiance[4]. In fact, it has been suggested in an example cited in Section 2 of Chapter 3 that one lasing line observed in a selenium plasma may be particularly narrow and radiating in superfluorescence, while others in the same ion radiate from ASE modes. Allen and Peters, in a classic series of papers[1,4−6], analyze the threshold conditions, output intensity, saturation, beam divergence, spatial coherence and spectral distribution for ASE. Much of their analysis was also verified with a 3.39-μm wavelength He-Ne laser, as described in the same papers.

In the ASE operating regime, the N_a^2 dependence of intensity does not hold. The net result of ASE operation can be an output beam from each end of a laser medium which can be extremely bright, moderately collimated (according to the aspect ratio a/L again), and with a fair degree of spatial coherence. Hence, for a very slender rod-like medium with a Fresnel number $\mathscr{F} \sim 1$, the emission will emerge in a single transverse mode. However, for a larger diameter a and $\mathscr{F} > 1$, a random superposition of $\sim \pi \mathscr{F}^2$ transverse modes will result. This is not dissimilar to a cavity laser with rather poor mode selection. The output will, however, still essentially be highly amplified gaussian noise, without any meaningful degree of temporal coherence. Hence, the medium acts overall like a mirrorless laser, with characteristics somewhere between a truly coherent cavity-type oscillator and an incoherent thermal source.

2.   GAIN FORMULATION AND REQUIREMENTS

2.1. Duration and Definitions

2.1.1. Self-Terminating vs. Continuous (CW)

There are essentially two modes of x-ray laser pumping. In the first (a), pumping is performed in a time less than that for filling the lower state. This filling may occur as a result of spontaneous radiative decay or perhaps by collisional or photon induced mixing from a nearby highly populated state.

Then, transient inversions are created in what is termed a "self-terminating" laser mode. These laser transitions may be between innershells in the atom or ion, with a high energy-conversion efficiency. However, spontaneous-decay times are in the 10^{-15} to 10^{-12} sec range for x-rays, with corresponding gain lengths limited by the speed of light to 0.3 to 300 μm, respectively. Such short lengths become quite impractical for finite-sized x-ray lasers, unless traveling-wave pumping is performed along the length in coincidence with the amplified wave. Such pumping is described and illustrated (Fig. 22) in Section 4.2 of this chapter (see Index, also).

Hence, for practical reasons, an alternate mode (b) of quasi-cw pumping is generally preferred. In this mode of operation, population inversions exist in non-equilibrium for the duration of pumping conditions. This mode of operation is possible because the lower-level population density depletes more rapidly than does the upper level.

Generally, pumping mode (a) is considered to have a higher overall efficiency than does (b). This is because the latter usually involves transitions between excited states with significantly lower transition energy for the same ion. Various possible and specific plasma pumping processes will be described in Chapters 3–5.

2.1.2. Basic Energy Levels and Labeling

In this section as well as in the detailed analyses in later chapters, we define the basic laser system as illustrated in Fig. 2. The upper and lower laser state population densities are designated by N_u and N_ℓ, respectively.

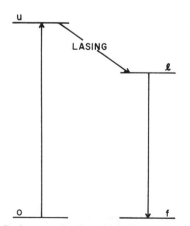

Fig. 2. Basic energy levels and labeling used throughout.

The final relaxation state density is labeled N_f. The laser wavelength is then referred-to as $\lambda_{u\ell}$. Also, we designate the density of the state from which the pumped electron originates as N_o. Where state "o" is an intermediate state in the entire process, the preceding "source" state is designated by the subscript "s".

2.2. Gain in ASE Systems

There are two major aims in this section. The first is to relate the increase in intensity in an elongated plasma, due to amplification of the spontaneous emission along the axis, to a gain coefficient G. With such relationships established, it is then possible to use the measured intensity enhancement to determine the gain coefficient. From that one can predict and scale the power output with further analysis and modeling.

The second goal here is to derive useful expressions for relating the gain coefficient to atomic processes. Such relationships can be used both to understand the observed features in a laser and to predict promising systems for further development.

2.2.1. Intensity Amplification Relations

2.2.1a. AMPLIFICATION OF AN INJECTED BEAM.   Gain in a photon beam making a single pass through an ideal uniform and homogeneous elongated medium of length L (illustrated in Fig. 1) and with a population inversion density is found from simple radiative transfer by integrating

$$\frac{dI}{dz} = GI. \tag{1}$$

The integration is performed over the length z from 0 to L to give the intensity I relative to the incident intensity I_o, i.e.,

$$\frac{I}{I_o} = e^{GL}. \tag{2}$$

Here G is the peak gain coefficient, i.e., at the center of the spectral line. (A subscript "c" on G is omitted for simplicity and clarity throughout the book). It is also the inverse of the mean free path for the photons in the medium. This is the quantity that is generally used to specify the amplification of intensity achieved. It is specified as an inverse length (e.g., cm^{-1}) for a given system. Sometimes, however, rather than through G, I/I_o is specified in terms of decibels (dB $= 10 \log_{10} I/I_o$) for a given length L, i.e., as dB/cm or dB/m.

For zero gain, Eq. (2) reduces simply to $I = I_o$. This relation [Eq. (2)] resembles that for absorption with a coefficient κ_c in a medium, i.e., $I/I_o = \exp(-\kappa_c L)$. Hence, G is sometimes referred to as a "negative-absorption" coefficient. A value of $G = 1$ is often considered a threshold for ASE. However, the usual goal for demonstrating x-ray ASE has been to achieve single-pass amplification by a factor of $e^5 = 150$, i.e., a gain-length product of $GL = 5$.

2.2.1b. ASE GAIN FROM PEAK LINE INTENSITIES. In a homogeneous ASE system, the intensity to be amplified is not fixed at an external value of I_o as depicted simply above. Rather it is the total spontaneous emission I_s distributed along the length of the gain region. Designating this spontaneous emission per unit length by J_s, the gain for an element dz (see Fig. 1) is given from Eq. (2) by $J_s \, dz \exp(Gz)$. The integrated intensity then becomes

$$I = \frac{J_s}{G}(e^{GL} - 1). \tag{3}$$

(This will be included in the summary of gain relations in Table 1 below.) For J_s and G fixed quantities, this output intensity scales with length as $(e^{GL} - 1)$. This formula can in principle be used for determining the peak gain coefficient G from measurements of peak intensity I versus length L (see Section 5 below). It is suitable for small gain, where in the limit e^{GL} approaches $1 + GL$ and I approaches $I_s = J_s L$.

When the total spontaneous emission integrated over the length of the medium is monitored (e.g., by an off-axis measurement) along with the amplified (axial) emission, it is convenient to use the ratio

$$\frac{I}{I_s} = \frac{\int_0^L J_s e^{Gz} \, dz}{\int_0^L J_s \, dz} = \frac{e^{GL} - 1}{GL}. \tag{4}$$

Again, for small gain, $e^{GL} \to 1 + GL$ and $I \to I_s$, i.e., simply spontaneous emission.

2.2.1c. ASE GAIN FROM INTEGRATED LINE INTENSITIES. The gain coefficient defined above is for line center, or peak. More often than not the xuv line will be too narrow to resolve the line profile instrumentally. Then it becomes necessary to include the frequency dependence due to the line shape. This can be done by expressing the intensity and the gain coefficient as the products $G\mathcal{S}(v)$. Here $\mathcal{S}(v)$ is the frequency (v) dependent shape function for the spectral line, normalized such that $\int \mathcal{S}(v) \, dv = 1$. The output intensity is then given by

$$I = \int_0^L dv \, \frac{J_s \mathscr{S}(v)}{G \mathscr{S}(v)} \, (e^{GL\mathscr{S}(v)} - 1) = \frac{J_s}{G} \, \mathscr{S} \, dv (e^{GL\mathscr{S}(v)} - 1). \tag{5}$$

This equation can be solved numerically. It can also be approximated analytically by applying either the method of steepest descents or a Taylor-series expansion as approximations to the exponential term[7]. The resulting analytical expression valid for any sharply peaked line profile, including Doppler, is[8]

$$I = \frac{J}{G} \, \frac{(e^{GL} - 1)^{3/2}}{(GLe^{GL})^{1/2}}. \tag{6}$$

This is also included in Table 1. This is the formula which is most generally and properly used for obtaining the peak ASE gain coefficient G from measured variations of integrated line intensity over length (e.g., see Section 5 below and Section 2 of Chapter 3).

For completeness, it is worth mentioning a simpler approximation that has been used when the gain-length product is $GL > 2$. This formula is obtained by the same mathematical methods but ignoring the -1 in Eq. (5). The result is

$$I = \frac{J_s}{G} \, \frac{e^{GL}}{(GL)^{1/2}}. \tag{7}$$

The numerical ("exact") solution to Eq. (5) as well as the approximations given in Eqs. (3), (6), and (7) are plotted[7] as $I/(J_s/G) = IG/J_s$ for comparison in Fig. 3. The deviations for each formula from the exact solution at high and low gain-length products discussed in the text above are apparent. For most applications, Eq. (6) is most adequate and has gained general acceptance.

Besides such a frequency dependence, other averages over time, space and angular spread should be included in data analyses as well as in modeling. Saturation effects occurring at high gain and discussed in Section 2.2.3. below are also omitted here.

2.2.2. Gain Coefficient Formulae

For net positive gain, the peak gain coefficient G must first and foremost exceed that for inherent line absorption in the laser medium. This coefficient is related to the cross-sections for stimulated emission σ_{stim} and resonance

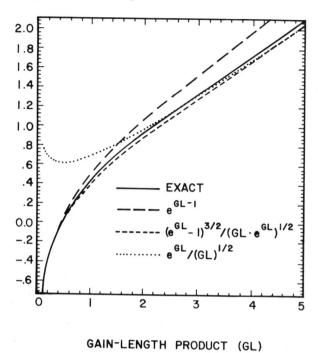

GAIN-LENGTH PRODUCT (GL)

Fig. 3. Length-dependent factors in Eqs. (3), (6) and (7) and the "exact" numerical solution to Eq. (5), versus the gain-length product GL. (From Ref. 7.)

absorption σ_{abs} and the upper and lower laser state population densities N_u, N_ℓ, respectively, by

$$G = N_u \sigma_{stim} - N_\ell \sigma_{abs} \doteq N_u \sigma_{stim} F. \tag{8}$$

Here

$$F \equiv 1 - \frac{N_\ell \sigma_{abs}}{N_u \sigma_{stim}} = 1 - \frac{N_\ell(g_u)}{N_u(g_\ell)} \tag{9}$$

is known as the inversion factor. The last step in Eq. (8) is possible because the cross-sections are related by corresponding statistical weights g_u and g_ℓ. The inversion factor F is made positive by efficient and preferential pumping of population into the upper-laser state. If N_u pumping is not so singular and the lower state density N_1 becomes appreciable (either by pumping or by transfer from N_u), the parameter F decreases and eventually becomes less than zero in equilibrium. This results in absorption, rather than amplification.

The upper state density N_u can be related to the initial state density N_0 for particular pumping modes. The latter is usually a ground state and more easily evaluated than the former. This connection is described in Section 2.2.2d, following a discussion of useful expressions for the cross-section σ_{stim}.

2.2.2a. GENERALIZED CROSS-SECTION FOR STIMULATED EMISSION. We first derive a general expression for the cross-section in terms of the spontaneous transition rate[4]. We begin with the geometry in Fig. 1, with $\mathcal{N}_v$ photons passing through the tube cross-section of area πa^2 at any point. For a population inversion density of N_u (assuming for simplicity that $F = 1$) in the medium, the net rate at which atoms or ions are induced to emit per unit volume at that point may be written $\mathcal{N}_v N_u \sigma_{\text{stim}}/\pi a^2$. Now, $\mathcal{N}_v \sigma_{\text{stim}}/\pi a^2$ is equivalent to $B_{u\ell} \rho_r(v)$, where $B_{u\ell}$ is the Einstein coefficient for induced emission. Also, $\rho_r(v) = Nhv/\pi a^2 c\, \Delta v$ is the radiation density, for a generalized line width Δv. Expressing $B_{u\ell}$ in terms of the spontaneous transition probability $A_{u\ell}$ through $B_{u\ell} = (c^3/8\pi hv^3)A_{u\ell}$, we arrive at the cross-section for stimulated emission

$$\sigma_{\text{stim}} = \frac{c^2}{8\pi v^2} \frac{1}{\Delta v} A_{u\ell} \doteq \frac{\lambda^4}{8\pi c\, \Delta \lambda} A_{u\ell} = \frac{\lambda^3}{8\pi c\, \Delta \lambda/\lambda} A_{u\ell}, \tag{10}$$

using $v = c/\lambda$ and $\Delta v = (c/\lambda^2)\, \Delta \lambda$ for the conversion to wavelength. [For brevity here (Section 2.2), we are ignoring the subscripts "$u\ell$" for frequency and wavelength. These will be used in following sections in the book to designate the laser transition, where the possibility of confusion with other transitions may arise.] The last form in Eq. (10) includes $\Delta \lambda/\lambda$, which is often specified instead of $\Delta \lambda$. This ratio is also convenient when converting to frequency or photon energy, because $\Delta \lambda/\lambda = \Delta v/v = \Delta(hv)/(hv)$. Equation (10) combines with Eq. (8) above to form two useful numerical expressions for the gain coefficient, listed in Table 1 as (10′) and (10″).

The frequency (and wavelength) dependence in Eq. (10) is exaggerated because of an additional inherent dependence in $A_{u\ell}$. We can derive an expression showing a weaker explicit wavelength dependence. For this we first relate the spontaneous transition rate $A_{u\ell}$ to the absorption oscillator strength $f_{\ell u}$ for the transition. The latter is often fairly constant throughout an isoelectronic sequence. This relation is

$$A_{u\ell} = \frac{8\pi^2 e^2 v^2}{mc^3} \frac{g_\ell}{g_u} f_{\ell u} \doteq \frac{8\pi^2 r_0 c}{\lambda^2} \frac{g_\ell}{g_u} f_{\ell u} \tag{11a}$$

$$\doteq \frac{6.6 \times 10^{15}}{\lambda^2} \frac{g_\ell}{g_u} f_{\ell u}, \tag{11b}$$

for λ in Angstrom units. Here $r_0 = e^2/mc^2 = 2.8 \times 10^{-13}$ cm is the classical electron radius. With this substitution, the cross-section in Eq. (10) then becomes

$$\sigma_{\text{stim}} = \frac{\pi r_0 c f_{\ell u}}{\Delta v} \frac{g_\ell}{g_u} = \frac{\pi r_0 f_{\ell u} \lambda^2}{\Delta \lambda} \frac{g_\ell}{g_u}$$

$$\doteq \frac{\pi r_0 f_{\ell u} \lambda}{\Delta \lambda / \lambda} \frac{g_\ell}{g_u}. \tag{12}$$

Hence, the cross-section expressed in this last manner has only a linear explicit dependence on wavelength, compared to a cubic dependence in Eq. (10). This also is included in Table 1 as (12′) and (12″) for numerical gain coefficient expressions, when combined with Eq. (8). In all of these formulae for cross-sections (and gain coefficients), there is an additional implicit and sometimes strong wavelength dependence of the pumping process leading to N_u. This will be illustrated in greater detail for specific cases in Chapters 3–5.

From Eq. (12), the cross-section has a value of $\sigma_{\text{stim}} \approx 1.5 \times 10^{-15}$ cm^2 at a wavelength of 182 Å in the soft-x-ray region, using as typical parameters a line-width ratio of $\Delta \lambda / \lambda = 3 \times 10^{-4}$, an absorption oscillator strength $f_{\ell u} = 0.64$, and a statistical-weight ratio $g_\ell / g_u = 4/9$. (These parameters were chosen for $n = 3$ to $n = 2$ transitions in one-electron hydrogenic ions, which has proven to be a particularly successful and easily analyzed lasing transition. It is used extensively as an example in Chapters 3–5.)

These basic gain relations show the importance of a large upper-state density and a strong, narrow laser line in achieving maximum gain, once a population inversion has ensured a positive gain through $F > 0$.

2.2.2b. SPECIFIC LINE SHAPES.   The cross-section for stimulated emission (and hence the gain coefficient) derived above is not yet specific in either the line profile or the width. For this further refinement, we can multiply Eq. (12) by a normalized shape function $\mathscr{S}_x(v)$. (Here the subscript x will be used to designate a particular line shape.) Likewise, we can specify $\Delta \lambda_x$ for a particular shape. Then the cross-section is given by

$$\sigma_{\text{stim}} = \frac{\pi r_0 f_{\ell u} \lambda^2}{\Delta \lambda_x} \frac{g_\ell}{g_u} \mathscr{S}_x(v). \tag{13}$$

For a Lorentzian profile typical of Stark effect line broadening associated with collisions of charged particles in the plasma, the shape function is given by[3,9]

$$\mathscr{S}_s(\lambda) = \frac{2}{\pi} \left[1 + \frac{4(\lambda - \lambda_c)^2}{\Delta \lambda_s} \right]^{-1}$$

$$= \frac{2}{\pi} = 0.64, \qquad \text{for } \lambda = \lambda_c, \tag{14}$$

at peak, where λ_c is the central wavelength. There is not a single convenient expression for the half width $\Delta\lambda_s$ of the Stark-broadened line in both Eqs. (13) and (14). If needed, the reader is referred to other books[10,11]. Fortunately, collisional broadening with such a Lorentzian profile is not usually as important as is Doppler broadening, for the plasma conditions most desirable for x-ray lasers. At the high densities where Stark broadening dominates for x-ray lines, often the densities are too large to maintain a population density inversion against collisional mixing processes[12,13].

Doppler broadening of the laser line is determined by the mostly random thermal motion of the lasing ions of mass M moving with a velocity determined by a kinetic temperature kT_i. A Gaussian-shaped profile is typical of Doppler-effect line broadening, and the line shape function is[3,14,15]

$$\mathscr{S}_d(\lambda) = \left[\frac{4\ln 2}{\pi}\right]^{1/2} \exp\left\{-4\ln 2\left[\frac{\lambda - \lambda_c}{\Delta\lambda_d}\right]^2\right\}$$

$$= \left[\frac{4\ln 2}{\pi}\right]^{1/2} = 0.94 \qquad \text{for } \lambda = \lambda_c. \tag{15}$$

The half width (full width at half intensity points) of the line is given by the ratio[16]

$$\frac{\Delta\lambda_d}{\lambda_d} = \frac{2(2\ln 2)^{1/2}}{c}\left[\frac{kT_i}{M}\right]^{1/2}$$

$$= 7.7 \times 10^{-15}\left(\frac{kT_i}{\mu}\right)^{1/2}. \tag{16}$$

Here kT_i represents the ion kinetic temperature in eV, M is the atomic mass, $\mu \approx 2Z$, and Z is the atomic mass number. Notice that for a typical value of $kT_i/\mu \approx 13$ in a high temperature transient plasma, $\Delta\lambda/\lambda \approx 3 \times 10^{-4}$. This value is adopted for some numerical relations in Table 1 and also used throughout Chapters 3–5.

Hence, the generalized peak cross-section and gain coefficient formulas are modified by factors of 0.64 for Stark and 0.94 for Doppler broadened lines. Therefore, a generalized formula such as given in Eq. (12) is quite adequate for most designs.

We can now write the cross-section for Doppler broadened lines in closed form, using Eqs. (10), (13), (15) and (16) as

$$(\sigma_{\text{stim}})_d = \frac{\lambda^3}{8\pi} A_{u\ell}\left[\frac{M}{2\pi kT_i}\right]^{1/2} \tag{17}$$

Table 1

Convenient Numerical Relations for Estimating Gain Coefficients[a]

General Relations	Equation

$$(I)_{\text{peak}} = \frac{J_s}{G}(e^{GL} - 1) \tag{3}$$

$$(I)_{\text{avg}} = \frac{J_s}{G} \frac{(e^{GL} - 1)^{3/2}}{(GLe^{GL})^{1/2}} \tag{6}$$

$$G = N_u \sigma_{\text{stim}} F \tag{8}$$

$$F = 1 - \frac{N_\ell g_u}{N_u g_\ell} \lessgtr 1 \tag{9}$$

$$N_u = \frac{N_o P_{\text{ou}}}{D_u} \tag{19}$$

General Line Broadening

$$G = 1.3 \times 10^{-36} \frac{\lambda^4 A_{u\ell}}{\Delta\lambda} N_u F \tag{10'}$$

$$G = 8.7 \times 10^{-21} \frac{\lambda^2 f_{\ell u}}{\Delta\lambda} \frac{g_\ell}{g_u} N_u F \tag{12'}$$

Fixed[b] $\Delta\lambda/\lambda = 3 \times 10^{-4}$

$$G = 4.3 \times 10^{-33} \lambda^3 A_{u\ell} N_u F \tag{10''}$$

$$G = 2.9 \times 10^{-17} \lambda f_{\ell u} \frac{g_\ell}{g_u} N_u F \tag{12''}$$

Doppler Broadening:

$$\frac{\Delta\lambda}{\lambda} = 7.7 \times 10^{-5} \left(\frac{kT}{\mu}\right)^{1/2} \tag{16}$$

$$G = 1.6 \times 10^{-32} A_{u\ell} \lambda^3 \left(\frac{\mu}{kT}\right)^{1/2} N_u F \tag{17'}$$

$$G = 1.1 \times 10^{-16} f_{\ell u} \lambda \frac{g_\ell}{g_u} \left(\frac{\mu}{kT}\right)^{1/2} N_u F \tag{18'}$$

Conversions

$$A_{ul} = 6.6 \times 10^{15} \left(\frac{g_\ell}{g_u}\right)\left(\frac{f_{\ell u}}{\lambda^2}\right); \tag{11}$$

$$v = \frac{c}{\lambda} = \frac{3 \times 10^{18}}{\lambda};$$

$$\Delta v = \left(\frac{c}{\lambda^2}\right)\Delta\lambda = \left(\frac{3 \times 10^{18}}{\lambda^2}\right)\Delta\lambda;$$

$$E_v = hv = \frac{hc}{\lambda} = \frac{1.24 \times 10^3}{\lambda};$$

$$\Delta E_v = \left(\frac{hc}{\lambda^2}\right)\Delta\lambda = \left(\frac{1.24 \times 10^3}{\lambda^2}\right)\Delta\lambda;$$

$$\frac{\Delta\lambda}{\lambda} = \frac{\Delta v}{v} = \frac{\Delta(E_v)}{(E_v)}.$$

[a] Units: λ [Å]; v [Hz]; E_v [eV]; G [cm^{-1}]; kT [eV].
[b] For high temperature plasmas near ionization equilibrium.

or, using Eq. (11), as

$$(\sigma_{\text{stim}})_d = \pi r_0 f_{\ell u} \lambda c \frac{g_\ell}{g_u} \left[\frac{M}{2\pi k T_i} \right]^{1/2}, \tag{18}$$

in terms of the transition probability and oscillator strength, respectively. These are repeated in numerical form as Eqs. (17') and (18') in Table 1 as convenient gain formulae.

It should be realized that other line broadening effects may enter. For example, plasma velocity effects on the line width due to ion-ion collisional narrowing as well as to hydrodynamic turbulence[17] may alter the gain by factors as great as 2–3. This was suggested in connection with some experimentally observed anomalies discussed in detail in Section 2 of Chapter 3.

2.2.2c. SUMMARY OF GAIN RELATIONS.   The gain coefficient relations are so important in estimating gain for planned systems that it is worth summarizing the most useful ones in a convenient numerical fashion, along with conversion relations. This is done in Table 1. The (unprimed) equation in the text to which each corresponds is also indicated. These expressions are used extensively in Chapters 3–5 in deriving gain coefficients for specific processes. All wavelengths here are for the laser transition. Hence, again, the subscript "uℓ" in $\lambda_{u\ell}$, used extensively later in the book, is omitted here for brevity.

2.2.2d. RELATING N_u TO N_0 WITH PUMPING.   The gain relations and associated scaling with wavelength in the previous sections are all expressed in terms of the upper state density N_u. This quantity depends on how the upper state is populated (pumped), as well as on the density of the initial source state o. Often the latter is a ground state for a particular ion, and its density N_0 is a better known quantity than N_u. There are also additional wavelength dependences that enter through the pumping process selected.

We may relate these quantities through the simple steady-state relation

$$N_u = N_0 \frac{P_{\text{ou}}}{D_u}. \tag{19}$$

Here P_{ou} is the pumping rate from level o to u and D_u is the total decay rate out of the upper state. For radiative decay, $D_u = \sum_i A_{ui}$, i.e., the sum of all the spontaneous emission rates out of the upper state. Other decay modes could include collisions and even autoionization, both of which are discussed in Section 2.4 below. This relation will be used extensively along with the relations derived above in arriving at gain coefficient formulae for specific pumping modes in Chapters 3 through 5.

2.2.2e. OTHER ABSORPTION LOSSES IN THE MEDIUM.   Besides line absorption, there are additional potential absorption and scattering processes for the laser beam which are not included in Eq. (8). One is photoionization (of cross-section σ_{pi}) in the medium at the laser frequency v and for a laser photon density N_v. There are three possible reasons for concern here. There may be:

(a) an additional direct loss term $-N_u\sigma_{pi}$ in Eq. (8),
(b) a loss of laser intensity from any level n at a rate $N_v\sigma_{pi}(n)c$, and/or
(c) significant depopulation of the upper state by photoionization.

From Eq. (23) of Chapter 4, Section 3, it is seen that peak photoionization cross-sections at an absorption edge are $\approx 10^{-19}$ cm^2. They also decrease generally as v^{-3}. Hence, they are many orders of magnitude smaller than σ_{stim}, estimated above to be typically $\sim 10^{-15}$ cm^2. Therefore concern (a) is usually not important. It follows from this that concern (b) should also not be a problem if (a) is not, since photoionization losses will be much less than line absorption, included in Eq. (8). Also, concern (c) will only occur when $N_v\sigma_{pi}(u)c > D_u$. This would be a saturation effect, and would occur at a much lower intensity for stimulated emission, as discussed in the next section. Hence, it is safe to conclude that photoionization effects in the laser medium will be negligible under normal conditions. Nevertheless, it is wise to compare the cross-sections and rates for any particular laser design, and if possible choose a laser wavelength below or very much above threshold for photoionization from the upper-laser level. (This is reemphasized in the summary of design criteria in Section 1.3 of Chapter 7.)

2.2.3. Saturation of Output at High Gain

As the gain-length product GL is increased in an ASE device, the output intensity does not continue to increase exponentially. Rather, it begins to saturate at a point where the photon-induced atomic processes compete with other processes in the amplifying atoms[3]. At this point, the population-inversion density becomes modified by the laser beam. At higher values of GL, the intensity rises linearly. Still there is large gain over the total length of the system. The onset of saturation represents the transition from the small-signal to the large-signal operating regime.

Such saturation has been studied[5,6,18] both theoretically and experimentally in the infrared spectral region. Saturation begins at the ends of the rod-like cylindrical laser medium, as a result of ASE built up from the opposite end. This continues such that there is only an increasingly narrow region of unsaturated gain in the central portion of the rod. There is also rapid line-narrowing, as the saturation threshold is approached with increasing gain and/or length. However, the line width is nearly constant or even rebroadens above threshold. The latter is referred to as "power broadening" in the laser field[18]. Sometimes in MOPA operation (see Section 1.4 of Chapter 1 and Index), a high-gain amplifier can drive a nearby oscillator to saturation, thereby turning it off prematurely.

Casperson[18] has shown that the saturation threshold occurs at $GL \approx$ 10–20. This has barely been reached experimentally for the xuv region. As the length is increased to approach this threshold, other density effects such as refraction (discussed in Section 2.4.5) serve to spread the beam.

2.2.4. Gain Without Reabsorption

It is also conceivable that gain may be obtainable without the line reabsorption that precipitates the need for a population-density inversion between the laser levels. If the stimulated emission spectral line differs in wavelength from the absorption line, the second term on the right in Eq. (8) can be ignored. Notice that this does not remove the overall requirement for a high non-thermal population density in the upper-laser state; it simply increases the gain by allowing F to approach unity. Nevertheless, it can make the difference between lasing and non-lasing in some situations.

There are at least three mechanisms by which this might be accomplished. In one rather specific innershell scheme described in detail in Section 3.3.4 of Chapter 5, an autoionization cascade serves to redistribute the bound electrons on a short time scale such that the K-shell emission occurs at a wavelength shifted from the absorption line. A second more general possibility that has received some theoretical attention[19,20] for xuv lasers involves a Doppler shift between the emitting and absorbing atoms. This is associated with additive recoil, both negative and positive, from the photon emission and absorption processes. A third suggestion[21,22] revolves around lasing from autoionizing states, and the different absorption and emission spectral shapes at autoionizing resonances. These rather novel concepts are admittedly difficult to implement and so far remain untested at any laser wavelength.

2.3. Gain in a (Lossy) Cavity

As usually perceived, a laser consists of an active medium wherein bound valence electrons are pumped to inverted population densities between atomic energy levels. The medium is placed within a cacity, for which the reflectance values of the mirrors may be designated by R_1 and R_2. For lasing in a homogeneous medium within such a cavity, the net amplification of an intensity I_0 to produce an output I following $\mathcal{N}$ reflections and $\mathcal{N} + 1$ passes over a path length L is given by[15,23]

$$\frac{I}{I_0} = R_1^{\mathcal{N}_c} R_2^{\mathcal{N}_c} \exp[(\mathcal{N}_c + 1)GL]. \tag{20}$$

Here again, G is the small-signal gain coefficient at line center with no saturation. For mirrorless ASE systems as described above, $\mathcal{N}_c = 0$ and Eq. (20) reduces to $I/I_0 = \exp(GL)$, similar to Eq. (2) in Section 2.2.

The obtainment of an efficient cavity is crucial to the ultimate development of a highly coherent and well-collimated x-ray laser of the quality associated with the infrared, visible and near-uv spectral regions. An added advantage of a cavity is the line narrowing afforded by multiple-pass amplification and the accumulated increase in overall gain, not possible with ASE[24]. Several possible cavity designs are discussed below, beginning with the more conventional type. The use of simple reflectors to verify gain in precursor cavity experiments is discussed in Section 5.5 below. Also, the channeling or waveguiding of the laser beam in an optimum manner for a particular cavity by tailored transverse electron-density gradients is discussed in Section 2.4.5c.

2.3.1. Cavity Configurations

2.3.1a. CONVENTIONAL CONFOCAL RESONATOR. A flat-mirror Fabry-Perot type etalon is the classical example usually taken for a laser cavity. However, a concave-mirror design is far more practical for alignment and overall efficiency. A particularly popular and efficient concave-cavity configuration is the confocal resonator. In this configuration, the focal length of each concave end-mirror is equal to the distance between the mirrors. Numerical wave analyses show that the diffraction losses in the medium are minimized in this arrangement[24]. In the most straightforward design, the laser emission is taken out through the front mirror which is either slightly transparent to the laser radiation[25] or has a small hole on-axis. Alternatively, the beam may be portioned out by dispersion from a grating placed in the cavity. This has been done with a grazing-incidence reflector at a wavelength of ~ 600 Å in Ne-like calcium[26].

An efficient cavity requires very high reflectance mirrors. For example, mirrors with 99% reflectivity are commonly used for visible and infrared lasers. While aluminum is highly reflecting in the visible region, the reflectance gradually decreases in the ultraviolet region as wavelengths become shorter and then drops rapidly at wavelengths shorter than 1200 Å. This and other possible normal-incidence reflectors, including the new layered synthetic microstructures, were discussed in Section 2.3 of Chapter 1. Such multilayer structures promise greatly enhanced reflectivity at xuv wavelengths. However, they tend to be extremely fragile and must be kept at considerable distance from the hot gaseous x-ray laser media. This again requires a long-pulse of synchronized pumping for multiple passes. Some results of survival tests in a plasma environment have been published[27]

2.3.1b. MULTIPLE-REFLECTION, GRAZING-INCIDENCE DESIGN. The reflectivity improves markedly for radiation incidence angles below the critical angle, i.e., at "grazing incidence" to the reflecting surface. This is discussed in Section 2.3.1b of Chapter 1. Also, the use of grazing incidence reflectors for soft x-ray imaging is discussed in Section 6.1.3 below. Circular cavities have been proposed, using grazing-incidence reflectors. A series of grazing-incidence mirrors would carry the laser beam in a fairly extended loop, returning through the lasant medium for multiple passes. Because light travels at a speed of approximately one foot per nanosecond, multiple-pass operation requires either a long-duration population inversion or a synchronized repetitively pulsed device. Such a cavity design is feasible but perhaps difficult to implement, at least at present.

2.3.1c. BRAGG-REFLECTION DESIGNS. A variation on such a circular-path cavity that is attractive for the harder x-ray region incorporates several crystals reflecting efficiently at the Bragg angle. There have been several proposals for such Bragg resonators.

A so-called puckered-ring design[28] is shown in Fig. 4. This resonator can be tuned to the laser wavelength. The losses at each crystal reflection are $<5\%$, but increase as elements are added. The difficulty in alignment is expected to be formidable. In another design[29] illustrated in Fig. 5, the resonator consists of pairs of parallel crystal planes. In this configuration, the polarization losses are less than for the puckered ring. However, accurate alignment dictates that each surface must be cut from a single crystal. Single crystals with alignments carefully fabricated have also been proposed[30] for resonators, as shown in Fig. 6. Finally, there is a suggestion[31] for the use of Borrmann crystals, which have an anomalously high transmission under certain circumstances (see Section 3 of Chapter 6).

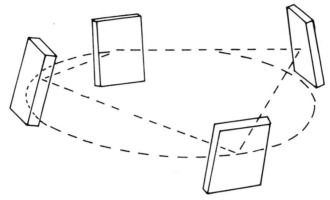

Fig. 4. Simplified drawing of the puckered-ring resonator. (From Ref. 28.)

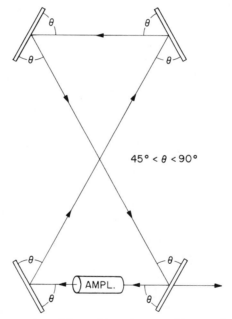

$45° < \theta < 90°$

Fig. 5. X-ray resonator and possible amplifier using several Bragg reflectors. (From Ref. 29.)

2.3.2. Distributed Feedback

Because of the extremely difficult problem of making and aligning x-ray resonance cavities, it has been suggested[32–35] that distributed feedback be
used instead. This would alleviate the need for sychronized operation for
gain over extended lengths. Distributed feedback is a technique already

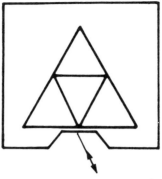

Fig. 6. X-ray resonator cut from a single crystal. The arrows show x-rays entering and exiting the resonator. (From Ref. 30.)

proven in the visible and infrared spectral regions. The feedback is distributed throughout the laser medium. Feedback is provided by Bragg scattering from a periodic variation of the refractive index in the medium or in the gain itself. The difficulties involved in applying distributed feedback to x-ray lasers center primarily around the transmission through the crystal. Also of concern is the distortion and destruction of the periodic structure of the crystal by the high pumping powers involved. Additional challenges and possible avenues to achieving x-ray lasing in crystals are discussed in Section 4.1 of Chapter 6.

2.4. Plasma Effects

The particle density in the plasma plays a key role in both promoting and limiting gain in x-ray laser systems. It is possible to make some preliminary estimates of density requirements and limitations. This is possible even before selecting a particular scheme and performing detailed analysis and modeling. Most of this section involves density effects. Plasma temperature effects are treated in Section 2.4.7 below.

2.4.1. Ion-Densities Required for High Gain

It is apparent from Section 2.2.2 that a large density (N_u) of lasing ions populated in the upper-laser-state is very important. It is also essential to have a small lower state density N_ℓ, for maximum inversion ($F \approx 1$). These two conditions are achieved either: (a) by preferential pumping of N_u, (b) by rapid depletion of N_ℓ through radiative decay, or (c) by a combination of these.

They are not mutually exclusive. For example, increasing the initial ion density N_o in order to increase N_u will most likely increase the final state density N_f. This in turn may increase N_ℓ if there is significant trapping of radiation on the ℓ-to-f depletion transition. A more quantitative discussion of these contrasting effects follows.

In assessing the initial ion density required, the general ASE goal has been to achieve single-pass amplification by a factor of at least $e^5 = 150$ times. This is equivalent to a gain-length product of $GL \geq N_u \sigma_{stim} FL = 5$ for a stimulated emission cross-section σ_{stim}. Using the cross-section expression given in Eq. (12) of Section 2.2.2 for a fixed value of $\Delta\lambda_{u\ell}/\lambda_{u\ell} = 3 \times 10^{-4}$ and taking account of a finite lower state density through the inversion factor $F \approx 0.3$, the requirement of $GL = 5$ leads to the criterion

$$N_u L = \frac{2 \times 10^{18}}{\lambda_{u\ell}} \quad \text{cm}^{-2}. \tag{21}$$

In arriving at this value we have again assumed the example of a hydrogenic-ion $n = 3$ to $n = 2$ lasing transition with a laser wavelength of $\lambda_{u\ell}$ in Angstrom units, a statistical weight ratio of $g_u/g_\ell = 9/4$, and an absorption oscillator

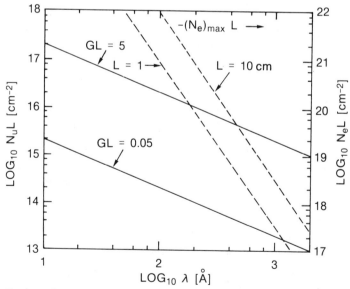

Fig. 7. Product of upper-state density and lasant length versus wavelength [Å] for two gain products: $GL = 5$ for ASE and $GL = 0.05$ for cavity operation. The corresponding product $N_e L$ is estimated as $\approx 10^4$ times the product GL, as indicated on the right side. Also plotted is $(N_e)_{max}L$ for $L = 1$ and 10 cm as determined by collisional mixing (without radiative trapping).

strength of $f_{\ell u} = 0.64$. Equation (21) is plotted in Fig. 7. Also plotted there for comparison is the equivalent condition for a typical cavity requirement of $GL = 0.05$.

We can relate this product $N_u L$ to that for the initial-ion density $N_o L$ using the relation $N_o = N_u(N_o/N_u)$. For the sake of the present argument, let us assume a somewhat conservative but typical value of $N_u/N_o \sim 3 \times 10^{-3}$. (In Chapters 3–5 we will discuss various pumping mechanisms and compare individual rates.) Then, Eq. (21) becomes

$$N_o L \approx \frac{6.7 \times 10^{20}}{\lambda_{u\ell}} L \quad cm^{-2}. \tag{22}$$

By a similar argument, it follows that the total ion density N_i (times the length) can be estimated from Eq. (22) to be

$$N_i L = N_o \left[\frac{N_i}{N_o}\right] L \approx \frac{2 \times 10^{21}}{\lambda_{u\ell}} L \quad cm^{-2}, \tag{23}$$

assuming $N_i/N_o \approx 3$.

Taking $N_e \approx 10 N_i$ for a moderate-Z element, the electron density is found from

$$N_e L = N_i \left[\frac{N_e}{N_i}\right] L \approx \frac{2 \times 10^{22}}{\lambda_{u\ell}} L \quad cm^{-2}. \tag{24}$$

Hence, in these rough estimates, $N_u L \approx 10^{-4} N_e L$; and the latter is included in Fig. 7 on the right ordinate for convenience. For example, ASE in C^{5+} ions at 182 Å requires for a $L = 1$ cm long lasant:

$$N_u \approx 1 \times 10^{16},$$

$$N_o \approx 4 \times 10^{18},$$

$$N_i \approx 1 \times 10^{19},$$

and

$$N_e \approx 1 \times 10^{20},$$

all in units of cm^{-3}. Here N_e is the free-electron density in the plasma. The latter two overall plasma conditions are consistent with those found in some recent experiments, to be described in Chapters 3–4.

At first glance the choice of length here appears to be arbitrary. Longer lengths indicate operation at lower densities for the same GL value. The importance of this will become more apparent as we discuss limits of the electron density next. The length cannot be increased arbitrarily, however. There is ultimately a practical limit to the total pumping power available. Also, another high density effect, namely refraction (discussed in Section 2.4.5b below), serves to limit the length. For this effect there is indeed a possible trade-off between density and length. Hence, the chosen gain conditions cannot simply be reached by increasing the density and/or length independently. Possible compromises will be discussed in Section 2.4.6, after describing the electron density and refraction effects in more detail.

2.4.2. Population Depletion Due to Electron Collisions

An indirect and possibly detrimental effect of increasing the total plasma-ion density (N_i) for increased gain occurs via the plasma electron density, as indicated above. In a quasi-neutral plasma, ions of net charge ζ are accompanied by an electron density $N_e = \zeta N_i$. In some x-ray laser schemes, the plasma electrons are used for collisional pumping of the population-density inversion. Here a large N_e is important to generate a large fractional inversion N_u/N_o, where again N_o is the initial-ion density.

2.4.2a. COLLISIONAL MIXING AND MAXIMUM ELECTRON DENSITY.   There is a definite and quite general limit placed on N_e by collisional mixing between levels u and ℓ of the laser transition. When the electron-collisional deexcitation rate from the upper- to lower-laser-level approaches the radiative decay rate of the latter, the lower-state density increases. Then, collisional equilibrium between the laser states is approached. This can be quantified quite readily by equating the deexcitation rate $N_e C_{u\ell}$ from Eq. (4) of Chapter 3 to the transition probability $A_{\ell f}$. It is convenient to substitute $A_{\ell f} = A_{u\ell}(A_{\ell f}/A_{u\ell})$, and thereby express both quantities in terms of the u–ℓ laser transition. Using Eq. (11) above for $A_{u\ell}$, the result is

$$(N_e)_{max}\left[\frac{1.6 \times 10^{-5} \langle g_{u\ell}\rangle f_{\ell u} g_\ell/g_u}{\Delta E_{u\ell}(kT_e)^{1/2}}\right] = 6.6 \times 10^{15}\left[\frac{A_{\ell f} f_{\ell u} g_\ell}{A_{u\ell}\lambda_{u\ell}^2 g_u}\right] \quad cm^{-3}, \quad (25)$$

for wavelengths $\lambda_{u\ell}$ in Angstrom units, and kT_e and $\Delta E_{u\ell}$ both in eV. Expressed in this way, the oscillator strengths and statistical weights conveniently cancel. Using $\Delta E_{u\ell} = 12400/\lambda_{u\ell}$, this can be written as

$$(N_e)_{max} = \frac{5.7 \times 10^{26}}{\lambda_{u\ell}^{3.5}}\frac{A_{\ell o}}{A_{u\ell}}\frac{(kT_e/\Delta E_{u\ell})^{1/2}}{\langle g_{u\ell}\rangle} \quad cm^{-3}. \quad (26)$$

This is still quite general.

As a more specific example, we assume that $\Delta n = 1$ for the laser transition and also that $kT_e \approx \Delta E_{u\ell}/2$. This limited temperature prevents excessive excitation of the lower state from the ground state. This point is discussed further in the modeling in Section 3.1.2 and Fig. 17 below, where this value corresponds to $kT = 0.1Z^2$ Ry for the hydrogenic $n = 3$ to $n = 2$ example. (The quantity Ry is the Rydberg energy equal to 13.5 eV.) This choice of temperature also gives a Gaunt factor[16] of $\langle g_{u\ell} \rangle = 0.2$ (see Fig. 2 of Chapter 3). In addition, we choose an average value of $A_{\ell f}/A_{u\ell} \approx 4$ (note that $A_{\ell f} \doteq A_{\ell o}$). These substitutions result in

$$(N_e)_{max} = \frac{8 \times 10^{27}}{\lambda_{u\ell}^{3.5}} \quad cm^{-3}. \tag{27}$$

Hence, it is possible to include the product $(N_e)_{max} L$ in Fig. 7, for both $L = 1$ and $L = 10$ cm. This indicates that for $L = 1$ cm, a maximum gain coefficient of 5 cm^{-1} is reached at a wavelength of 200 Å. This agrees with experiments (see Chapters 3–5). A reduced GL product and overall gain occurs at longer wavelengths unless L is increased. In other words, longer-wavelength lasers require longer lengths for the same overall gain, as the density decreases.

For the example of lasing on the $n = 3$ to 2 Balmer-α transition in a hydrogenic ion of nuclear charge Z, $\lambda_{u\ell}$ can be approximated by $6562/Z^2$ in Angstrom units. Then Eq. (27) gives a conservative z-scaling relation

$$(N_e)_{max} = 3 \times 10^{14} Z^7 \quad cm^{-3}. \tag{28}$$

For $Z = 6$ and $\lambda_{u\ell}$ 182 Å, $(N_e)_{max} = 8 \times 10^{19}$ cm^{-3}. Again, this is in agreement with a value derived in greater detail in the example presented in Section 3.1.2 of this Chapter. This value will be used extensively in specific examples in Chapters 3–5.

Fortunately, the limit $(N_e)_{max}$ increases much more rapidly ($\sim Z^7$) than does the photon energy ($\sim Z^2$). Hence, as seen in Fig. 7, much higher densities and gain should be possible at shorter wavelengths and shorter lengths with higher-Z plasmas, barring other pumping and opacity limitations to be discussed later.

2.4.2b. OTHER ELECTRON-COLLISIONAL DEPLETION EFFECTS. Besides the population mixing between the laser levels described above, a population density depletion may result from ionization. Of the two effects, the former is generally the more important because of a larger cross-section. Additional electron-impact broadening associated with charged-particle collisions, most notably due to the Stark effect, was discussed in Section 2.2.2b.

2.4.3. Opacity, Size, and Trapping Limitations

2.4.3a. STATIC CASE.   The effect of radiative reabsorption in the lower-laser-level depletion transition on the population of this level, and hence on the population inversion, is of vital importance to achieving and maintaining the gains indicated in Fig. 7. The magnitude of the effect can be evaluated by modifying the transition probability with an "escape factor" $G(\tau_c)$. Then, $A_{\ell f}$ becomes $G(\tau_c)A_{\ell f}$ (see Section 3.1 of this Chapter and the example therein). This escape factor is unity for optically thin lines and decreases rapidly as the opacity τ_c at line center exceeds unity (see Fig. 8).

The opacity at line center is defined by $\tau_c = \kappa_c d$ (κ_c being the corresponding absorption coefficient). These quantities determine the amount of absorption occurring in a path length d in a static medium through the Lambert-Beer Law

$$\frac{I}{I_0} = \exp(-\tau_c) = \exp(-\sigma_c N_f d). \tag{29}$$

Thus, the intensity I_0 is reduced to I after passing through a layer of thickness d with N_f absorbers per cm^3, each with a total absorption cross-section σ_c. For a Doppler-broadened spectral line in a static plasma, the opacity τ_c at the central wavelength λ_c in a static plasma of thickness d may be obtained

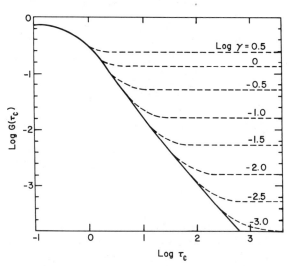

Fig. 8.   Escape factors $G(\tau_c)$ versus optical depth τ_c at line center for Doppler broadened transitions. The bold lower envelope represents the static case, i.e., without flow. The dashed extensions represent dynamic flow with velocity gradients. The labels designate the gradient of the velocity respect to τ_c, normalized to the thermal velocity.

from[10]

$$\tau_c = r_o \left(\frac{\pi M c^2}{2kT}\right)^{1/2} N_f \lambda_c f_{f\ell} d, \tag{30}$$

where e and r_o are the charge and classical radius of an electron, $f_{f\ell}$ is the absorption oscillator strength, M is the atomic mass, λ_c is the center wavelength, N_f is the absorbing particle density, and kT is the kinetic temperature of the ion (k being the Boltzmann constant). This reduces to[16]

$$\tau_c = 1.1 \times 10^{-16} \lambda_c N_f df_{f\ell} \left(\frac{\mu}{kT_i}\right)^{1/2}, \tag{31}$$

with $\mu \approx 2Z$ the atomic mass number, and for kT in eV, N_f in cm^{-3}, λ_c in Å, and d in cm. Obviously the opacity can be reduced by shortening the depth of the plasma. This is a major reason for using small diameter plasmas in most laser schemes.

Some useful calculated escape factors $G(\tau_c)$ are plotted in Fig. 8 versus τ_c. The heavy, solid lower envelope is computed[36-38] for Doppler-broadened lines in a non-flowing plasma. For $\tau_c < 0.5$, this can be approximated by $G(\tau_c) \approx [1 - \exp(-\tau_c)]/\tau_c$. Also, for $\tau_c > 3$, $G(\tau_c) \approx (\pi \tau_c^2 \ln t_o)^{-1/2}$ can be used[36]. The decrease in the $G(\tau_c)$ envelope with increasing optical depth is not nearly as rapid for Lorentzian line profiles, as calculated by Hummer and Rybicki[38]. These authors also calculate and plot escape factors for Voigt convolutions of Doppler and Lorentzian components in various proportions.

As an example of opacity effects in an x-ray laser static plasma, consider again lasing on a $n = 3$ to $n = 2$ transition at a wavelength of 182 Å in hydrogenic C^{5+} ions ($\mu \approx 12$). A typical plasma temperature for an abundance of excited ions of this species would be[39] 100 eV. Depletion from the $n = 2$ level then occurs on a Lyman-α $n = 2$ to 1 transition, for which the wavelength is $\lambda_{\ell f} = 33.7$ Å and the oscillator strength is[40] $f_{f\ell} = 0.42$. The optical depth from Eq. (31) becomes $\tau_c = 5.4 \times 10^{-16} N_f d$. This opacity reaches unity at a typical x-ray laser ground state density of $N_f = 10^{18}$ cm^{-3} for a depth (diameter) of $d \approx 20\mu$m. The size decreases rapidly for shorter-wavelength lasers operating at higher density. Hence, the currently popular use of small vaporized-fiber plasmas described in Section 2 of Chapter 4.

2.4.3b. EXPANDING PLASMA CASE. The dashed extensions towards larger τ_c in Fig. 8 represent modifications to the above static formulism for a flowing plasma with a large and constant velocity gradient. The labels refer to values of $\gamma = \Delta(v/v_{th})/\Delta\tau_c$. For this, the velocity is measured in units of the thermal velocity v_{th} [$= (2kT/M)^{1/2}$] and the gradient is taken with respect to the optical depth. This potential reduction in radiative trapping for large velocity gradients results from a decrease in optical depth at line center associated with the added Doppler shift due to the velocity gradient.

This modification was originally suggested for astronomical purposes[37,38,41] and more recently shown to be relevant to rapidly expanding laser-produced plasmas[42-44]. As an example relevant to x-ray laser plasma conditions, an expanding laser-produced plasma with a thermal velocity of $v_{th} = 10^6$ cm/sec might undergo a change in streaming velocity of $\Delta v = 10^7$ cm/sec in a distance over which the optical depth for the reasonance line varied by $\Delta \tau_c = 100$. Hence, $\Delta(v/v_{th})/\Delta \tau_c = 0.1$ and there should be a significantly enhanced escape factor according to Fig. 8. This then translates into a reduced degree of radiative trapping. This reduction in the effective optical depth can also be derived from the expression for the optical depth given in Eq. (30) for the static case, simply by replacing the ratio of the thermal velocity to the plasma depth (d) by the velocity gradient[41,43].

2.4.4. Detrapping: Reducing the Lower-Level Density

A reduction in line absorption described above for large velocity gradients can in principle greatly facilitate the use of larger-diameter plasmas. This is important for large x-ray gain without degradation in population inversion through an enhanced lower-level population. Hence it is recognized as one form of "detrapping" for x-ray lasers. Other possible methods, besides the Doppler shifting described above, include reducing the lower-laser-level population buildup, creating an alternative lower-level decay mode, or absorbing the decaying (trapped) radiation in a separate medium. These are described in the following subsections.

2.4.4a. AUXILIARY PUMPOUT BY PHOTOEXCITATION.   In this approach, the lower-laser-level population density is reduced by photoexcitation. Consider first the case where excitation takes place into a high-lying bound state. Such an excited state can undergo additional collisional mixing, ionization, and alternate decay. Hence, cascade directly into the laser system is reduced by branching through the multiple decay channels. This is illustrated by the bold, solid arrows originating on level h in Fig. 9. In this figure, the initial photoexcitation is shown taking place from level 1 to level h. This process does not significantly enhance the gain through direct upper-level population replenishment as in the case to be described in the next paragraph. However, it does serve to indirectly enhance the inversion through lower-level deple-

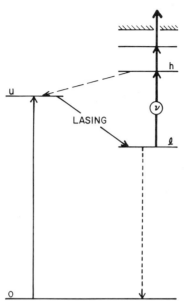

Fig. 9. Removal of lower-laser-level electrons by photoexcitation and photoionization to close levels, from which repopulation of the inversion can occur.

tion. The cross-section for direct photoionization from level l is much smaller than that for excitation. Therefore, any complete removal of the electron from the ion (ionization) would probably proceed in a stepwise process of excitation followed by ionization, as indicated.

If such photo-excitation takes place into a level above but still near the upper-laser-level, the electrons can cascade into this level. Then, the gain is increased through population replenishment. This possibility is included in Fig. 9, where photoexcitation from level ℓ to level h decays directly to the upper-laser-level u, shown as a dashed transition. Such preferential photo-excitation is best accomplished by line radiation at a photon energy matched with that of the absorbing transition.

Such resonant photoexcitation is analyzed in detail in Section 3 of Chapter 3, where the emphasis is placed on selective pumping from the ground state. One example described and included in Table 6 there features the pumping of $n = 2$ electrons into an $n = 4$ upper-laser state in hydrogenic ions of nuclear charge $2Z_p$. This resonant pumping is done with matching photons radiated by a Lyman-α $n = 2$–1 transition in a hydrogenic ion of nuclear charge Z_p. Lasing then occurs mostly by $n = 4$ to $n = 3$ transitions. This same principle can be applied to detrap the $n = 2$ lower-laser-level in a $n = 3$

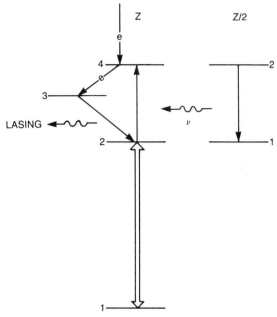

Fig. 10. Energy level diagram for photo-excitation enhancement of gain and decrease of trapping through lower level ($n = 2$) depletion in hydrogenic ions.

to 2 Balmer-α laser transition[45]. This is detailed in Fig. 10, which includes the additional option of pumping $n = 2$ electrons into an $n = 6$ state by Lyman-β $n = 3$ to 1 transitions in the pumping ion. This 3–2 laser transition is very popular, particularly for electron recombination pumping followed by cascading as described in Chapter 4. This pumping and cascading is indicated by the electron path labeled "e" in Fig. 10.

The necessary condition for this detrapping to be effective is that

$$N_2(N_\nu \sigma_{24} c) \geq N_3 A_{32}, \tag{32}$$

for the lasing ion, i.e., that the $n = 2$ pumpout rate for a photon density N_ν must exceed the filling rate from the $n = 3$ level. Using this along with Eq. (30) of Chapter 3 and assuming that $N_3/N_2 = g_3/g_2 = 9/4$, we can derive a necessary photon density of

$$N_\nu = \frac{9\pi}{2} \frac{1}{\lambda_{42}^3} \frac{A_{32}}{A_{42}} \left[\frac{\Delta \nu}{\nu} \right]_{42}. \tag{33}$$

For a C^{5+} ion lasing on the $n = 3$ to 2 Balmer-α transition and detrapped by a $n = 2$ to 1 Lyman-α photon from a Li^{2+} ion (n represents the principal quantum number), this photon density is

$$N_\nu(Z_p = 3) = 9 \times 10^{15} \quad \text{photons/cm}^3, \tag{34}$$

where we assume $\Delta\nu/\nu = 3 \times 10^{-4}$ again and hydrogenic parameters[40]. Such a photon density can be obtained at the center of an optically thick spectral line at a brightness temperature of $kT_B = 300$ eV or a pumping power of 5 MW for this particular combination[45]. For another combination, namely Mg^{11+} lasing ions pumped by Lyman-α from C^{5+} ions, the required brightness temperature is estimated to be $kT_B = 1.2$ keV and the pumping power 1.2 GW. These values are not unreasonable by present standards.

2.4.4b. AUTOIONIZING LOWER-LASER-LEVEL. A novel method of replacing lower-level radiative decay and trapping by an alternate process is to promote lasing on a transition in which this level autoionizes at a higher rate than radiative decay. Hence, the energy stored in the lower-laser-state is released as a free electron rather than as a photon[46–48]. This is illustrated in Fig. 11. The upper-laser state must be relatively stable against autoionization, which is possible for certain levels according to selection rules. Also, the rate

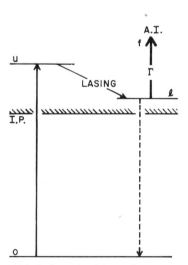

Fig. 11. Lasing on ions in which the lower level decays by autoionization, avoiding lower-level radiative trapping.

of autoionization scales as n^{-3}, n being the principle quantum number, which favors higher-level stability. Selective pumping of the doubly excited upper level by photoexcitation or even dielectronic recombination must be invoked, as discussed in Section 3 of Chapter 3 and Section 5 of Chapter 4, respectively. Photoexcitation requires the availability of an intense pumping line from another ion at a matching energy.

A promising system for this process is based on the three-electron Li-like isoelectronic sequence. Promotion of a single is electron can result in a $1s2\ell3\ell'$ upper configuration above the ionization limit. (Here ℓ and ℓ' represent s, p and s, p, d states, respectively.) Certain terms in such configurations are both relatively slow to autoionize and also couple radiatively in potential x-ray lasing transitions to $1s2\ell2\ell'$ lower-laser-levels. These states in turn autoionize at a rapid rate compared to radiative decay. Wavelengths of such 3–2 transitions are plotted in Fig. 12. Such transitions involving autoionizing states can have large natural widths, which reduces the gain [see Eqs. (8)–(10) above]. Hence, it is important to select transitions with lower-level autoionizing rates not so large that Doppler widths are exceeded.

All of these criteria have been considered[46,47] and some particularly promising candidates have been identified. This survey was based on available

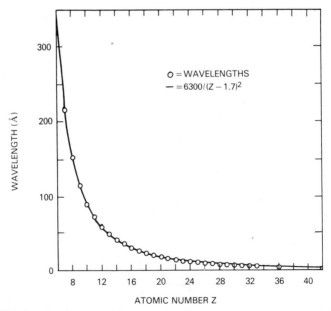

Fig. 12. Wavelengths of 3–2 transitions in the autoionizing continuum versus atomic number Z. An approximate-fit formula is shown. (From Ref. 47.)

Table 2

Combinations for Lasing with an Autoionizing Lower Level in Li-Like Ions[47]

Z_{pump}	λ_p [Å]	Z_{abs}	λ [Å]	$\Delta\lambda/\lambda\,(\times 10^4)$	Transition
Hydrogenic (2p-1s) Pump					
9	14.9823	9	14.9822	0.13	$1s^22s$-$1s2s3p$
		9	14.9825	0.33	$1s^22s$-$1s2s3p$
9	14.9877	9	14.9822	3.9	$1s^22s$-$1s2s3p$
		9	14.9825	3.7	$1s^22s$-$1s2s3p$
9	14.8823	9	14.990	5.3	$1s^22p$-$1s2p3p$
9	14.9877	9	14.990	1.3	$1s^22p$-$1s2p3p$
21	2.7360	20	2.7333	9.9	$1s^22s$-$1s2s3p$
He-Like (2p-1s) Pump:					
16	5.0389	15	5.0380	1.0	$1s^22p$-$1s2p3p$
			5.0409	4.8	$1s^22p$-$1s2p3p$
28	1.5885	26	1.5884	0.63	$1s^22p$-$1s2p3p$
			1.5858	17	$1s^22s$-$1s2s3p$
			1.5869	10	$1s^22s$-$1s2s3p$
29	1.4777	27	1.4773	2.7	$1s^22p$-$1s2s3d$

theoretical values of photon energy matches. Some of the findings are re-produced from Ref. 47 in Table 2 here for two different intense pumping lines, namely from the 2p–1s resonance transitions in hydrogenic and helium-like ions. A particularly interesting candidate system shown involves hydrogenic fluorine pumping lithium-like fluorine[48], i.e., both pumping and lasing ions of the same element. This could be an promising candidate for gaseous-discharge (e.g., z-pinch) plasmas (Section 4.2.2b below).

2.4.4c. ABSORPTION IN ANOTHER ELEMENT.   A third method of reducing the radiative trapping effect involves the use of a separate absorbing-ion admix-ture. In this scheme, the photons subject to trapping are absorbed and used to photoexcite a ground-state electron in an ion of a different species. A very-nearly exact energy match is necessary between the trapped photon and the absorption transition. This therefore relates closely to the photo-excita-tion concepts developed in Section 3 of Chapter 3. The addition of a second ionic component tends to "dilute" the available gain. This is because both species contribute free electrons to the plasmas, and there is a limit on the maximum electron density tolerable. Hence, the corresponding number of lasing ions must be decreased by at least a factor of two to allow for the detrapping ions. Also, it is important in this method that the absorbing ion

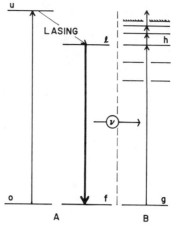

Fig. 13. Lower-laser-level detrapping by photo-excitation and ionization in an admixture ion.

have a large opacity (i.e., high absorption), for efficient coupling. Therefore, a dynamic flow environment such as described in Section 2.4.3b would not be desirable, because it would increase the escape factor and decrease the absorption.

If photoexcitation of the absorbing ion takes place into a sufficiently high-lying level, the excited electron decays through multiple collisional and radiative channels and the energy is ultimately released over a broad spectrum[49]. This is illustrated in Fig. 13, where the excited level is labeled "h" and further excitation is shown to higher levels. Such a system must be designed very carefully. The electron density is already high in a lasing environment. Also, the density of absorbing ions (and accompanying electrons) must be sufficiently high to compete with self-absorption (trapping) in the lasing ions. On the other hand, if the absorbing ion and the electron densities are so large that collisional equilibrium exists among the excited levels, this ion will re-radiate strongly at the same frequency as that absorbed. Then there occurs the possibility that the detrapping effect may be nullified. Again, photoionization into the continuum from the initial (ground) state is less likely than excitation followed by ionization in a multistep process, at least for resonance absorption. The equations necessary for detailed analysis are developed in Section 3 of Chapter 3.

In a variation on this absorber approach to detrapping, the discrete photo-excited state lies above the ionization limit. It then autoionizes at a rate much larger than that for radiative decay. Hence, the photon energy from laser depletion will be transferred finally to the kinetic energy of a free electron[46]. This is illustrated in Fig. 14. An advantage to this method over

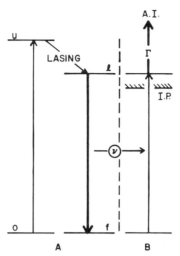

Fig. 14. Lower-laser-level detrapping by population of an autoionizing level through photo-excitation, in an admixture ion.

that described in the previous paragraph is that the natural width of the absorption line can be much larger, because of a large autoionization rate. This makes energy matching much easier. Also, some Doppler shifting associated with relative flowing between ions could be tolerated, if necessary. Numerous possible combinations of lasing followed by absorption-to and autoionizing-on $n = 2$ and $n = 3$ double excited states in two-and three-electron ions are listed in Ref. 46. Certain combinations with reasonably favorable coincidences (i.e., $\Delta\lambda/\lambda \leq 20 \times 10^{-4}$) between the wavelength λ_{depl} of the depletion transition from the lower-laser-level and the wavelength λ_{abs} of the absorption transition to the autoionizing level are listed in Table 3 here. The parameter n'' refers to the quantum number of the autoionizing electron.

2.4.5. *Uniformity, Homogeneity and Refraction Losses*

There are potentially adverse density effects on the x-ray laser output that are independent of the gain coefficient and have to do with the plasma uniformity. This concerns both axial and transverse directions.

2.4.5a. AXIAL DISTRIBUTION OF AMPLICATION.   So far, it has been assumed in summing the simple relation $\exp(Gz)$ along a length L for net gain that the rod-like plasma is straight, uniform and homogeneous in density and pumping conditions along the axis. This is the ideal radiative transfer condition.

Table 3

Combinations for Detrapping Absorption to Autoionizing Levels[46]

Species (Depletion)	λ depl [Å]	Species (Absorb)	λ_{abs} [Å]	$\Delta\lambda/\lambda^a$ (× 10⁴)
Li-Like $n'' = 2$ Absorber				
Si(Li-) [3p-2s]	40.911	C(Li-) [1s²2s-1s2s2p]	40.875	8.8
K(Ne-) [3s-2p]	41.541	C(Li-) [1s²2p-1s2p²]	41.563	5.4
K(Li-) [3d-2p]	22.02	[1s²2s-1s2s2p]	22.020	0
		O(Li-)		
	22.16	[1s²2p-1s2p²]	22.12	2.7
Ne(He-) [3p-1s]	11.5466	Na(Li-) [1s²2p-1s2s²]	11.537	6.7
Zn(Ne-) [3s-2p]	11.51	Na(Li-) [1s²2p-1s2s²]	11.534	17
He-Like $n'' = 2$ Absorber				
Na(Li-) [3d-2p]	77.764	Be(He-) [1s2s-2s2p]	77.829	8.3
N(He-) [3p-1s]	24.898	N(He-) [1s2p-2p²]	24.881	6.8
Ca(Ne-) [3s-2p]	15.157	F(He-) [1s2s-2s2p]	15.157	8
Cu(Ne-) [3s-2p]	12.558	Ne(He-) [1s2p-2s²]	12.553	4
Li-Like $n'' = 3$ Absorber				
Cl(Li-) [3p-2s]	26.67	N(Li-) [1s²2p-1s2s3s]	26.690	
Ti(Ne-) [3s-2p]	26.641			18
Ca(Li-) [3d-2p]	19.64	[1s²2p-1s2s3d]	19.615	13
		O(Li-)		
	19.79	[1s²2p-1s2s3s]	19.794	2
F(H-) [2p-1s]	14.982	[1s²2s-1s2s2p]	14.964	12
		F(Li-)		
	14.988	[1s²2p-1s2p3p]	14.990	1.3
Ca(Ne-) [3s-2p]	15.169	F(Li-) [1s²2p-1s2s3d]	15.138	20
Ne(H-) [2p-1s]	12.132	Ne(Li-) [1s²2p-1s2s3s]	12.121	9.2
He-Like $n'' = 3$ Absorber				
O(He-) [2p-1s]	21.620	N(He-) [1s2s-2s3p]	21.611	4.2

^{a} $\Delta\lambda = |\lambda_{depl} - \lambda_{abs}|$.

However, this condition is unlikely ever to be found in an actual experiment. This is particularly true at the high plasma densities required for ASE operation.

Plasmas produced by line-focused laser beams with high beam quality impinging onto condensed targets most nearly meet these conditions. An example is illustrated in the isodensity map of a portion of an x-ray pinhole photograph reproduced in Fig. 15. This was obtained with a 1.05-μm wavelength, 100–300 J, 2-ns FWHM, $\leq$800-ps-risetime Nd:glass laser. The beam was focused with a combination spherical and cylindrical lens system to a line of $\lesssim$ 100 μm width and 6–30 mm length on a thin formvar-backed copper

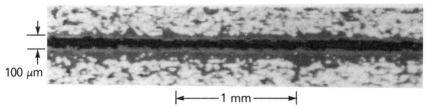

100 μm

|←——————1 mm——————→|

Fig. 15. Isodensity map of a portion of an x-ray pinhole photograph from a laser-produced line plasma (From Ref. 50.)

film target[50]. The macroscopic uniformity shown is considered excellent overall. This can be attributed to excellent beam and surface quality, as well as to a short rise-time and a low pre-pulse on the driving laser pulse. Still, there exist microscopic structures associated with filamentation and turbulence effects that can influence gain (see Section 2 of Chapter 3).

In contrast to the generally satisfactory conditions in laser-produced plasmas, x-ray pinhole photographs of elongated plasmas generated by pulsed electrical discharges in gases (or exploding solids) have been non-collimated and fragmented. An example is shown in Fig. 16 from an aluminum plasma[51]. This is typical of z-pinch plasmas and is generally associated with inherent "sausage" instabilities.

2.4.5b. TRANSVERSE ELECTRON DISTRIBUTION AND REFRACTION. The transport of the laser radiation down the length of the x-ray laser medium is subject to refraction out of the narrow gain channel. This results from transverse electron-density gradients. The effect is particularly a problem for electron densities $N_e \gtrsim 10^{20}$ cm^{-3} in plasmas of width $\lesssim 100$ μm, such as those produced by transverse laser irradiation of condensed targets. Excess refraction results in the premature loss of photons from the channel of amplification. It also can reduce collimation and coherence in the beam.

This problem has been analyzed thoroughly by Chirkov[52]. He considered first the case of a linear change in density with distance from the irradiated surface. This he considered to be typical of thin film as well as solid targets. Chirkov also analyzed a quadratic decay of electron density with distance from the focal spot (line). With this he was particularly modeling expansion

Fig. 16. X-ray pinhole photograph of a pulsed-power-driven (exploding-wire) z-pinch plasma. (Adapted from Ref. 51.)

in a direction parallel to the target surface. His overall results agree quite well with the simple analysis which follows.

The radial deflection angle θ_r of the refracted x-ray laser beam passing through a plasma of length L and bent toward the incoming driving beam can be estimated from

$$\theta_r = \frac{r_0}{2\pi} L\lambda_{\mu\ell}^2 \frac{dN_e}{dr}. \tag{35}$$

Here it is assumed that dN_e/dr is constant with an average value over the length L. Also, r_0 is again the classical electron radius and $\lambda_{u\ell}$ is the x-ray laser wavelength. At first glance, the strong scaling with wavelength appears to favor shorter wavelength lasers. However, the electron density can be expected to scale up with shorter wavelengths, perhaps as rapidly as $\lambda_{u\ell}^{-3.5}$ along a hydrogenic isoelectronic sequence (see Section 2.4.2a). That can counteract the advantage of a decrease in refraction with shorter wavelengths[52].

Using an average value for a smooth gradient over the radius r, Eq. (35) may be rewritten as

$$\theta_r = \frac{r_0}{2\pi} \lambda_{u\ell}^2 \frac{\Delta N_e}{N_e} \frac{N_e}{\Delta r} L. \tag{36}$$

Hence, the refraction angle scales as the electron density, if the other parameters are known. A typical value for the fractional change in density of $\Delta N_e/N_e \approx 0.3$ over the radial scale length Δr is adopted from a selenium x-ray laser experiment[53] discussed in Section 2 of Chapter 3. With this approximation, Eq. (36) then can be rewritten conveniently as

$$\theta_r = 1.3 \times 10^{-30} \lambda_{u\ell}^2 \frac{N_e L}{\Delta r}, \tag{37}$$

for a laser wavelength $\lambda_{u\ell}^2$ in Angstrom units. As an example, we can assume $\lambda_{u\ell} = 200$ Å, a radial scale length of 100 μm, and a maximum electron density product with length of $N_e L = 10^{20}$ cm^{-2} corresponding to a gain produce of $GL = 5$ and a length of $L = 1$ cm taken from Fig. 7. We then derive a refractive spread angle of approximately $\theta_r = 1$ mrad. This would be an excellent degree of collimation if achieved (see Section 2.5 of Chapter 7).

Continuing with the selenium x-ray laser example initially conducted under these same conditions (except for an electron density two to three times larger), a refractive deflection of the x-ray laser beam was not measurable[54]. However, in similar experiments with the length gradually increased from 1 to 4.4 cm, there was a lack of increase of axial gain which was attributed to a refractive fanning of the beam. For this length, the deflection is predicted from Eq. (37) to reach 8 mrad. It was later measured[55] to be ~ 10 mrad, in good agreement with this simple model.

Hence, electron densities above 10^{20} cm^{-3} lead to severe beam spreading due to refraction, for reasonable lengths of a few centimeters. Therefore, if these densities are essential, it is important to develop techniques to minimize this effect or at least compensate for it. Properly mastered, such lensing techniques could serve to channel the laser beam continuously along the axis of amplification. This could result in collimation in waveguide fashion, for optimized operation within cavities[52]. There have been various suggestions for contouring the plasma with an increasing density from the center outwards along the transverse dimension. Both slotted [52,56], humped[52], and layered[57] target designs for transverse laser drivers have been suggested for this purpose.

2.4.5c. BEAM GUIDING.   In an article by Fill[58] it is suggested that the observed beam displacement with longer lengths may be due to inhomogeneities, rather than refraction. The argument is put forth that the beam may be guided along the length by the gain profile when the maximum symmetry axis, along with that of the index of refraction, is on the cylindrical axis. As of 1989 this interesting alternative remains a matter of conjecture.

2.4.6. Summary and Tradeoffs: Density Effects

Summarizing the effects of plasma density on net gain, it is first of all clear that as high a density is possible is needed for ASE operation without an efficient cavity. On the other hand, too high a density can lead to destruction of population inversion through collisional mixing, enhanced line broadening, and opacity, all of which decrease the gain. Hence, ideally one would design for moderate density and achieve high gain by intense pumping over lengths measured in meters. This implies sustained pumping provided by long duration pulses (approximately a nanosecond for each foot of length).

However, available pumping-power density is limited for very short wavelengths, as discussed in Section 2.5. One way to approach this problem and improve the overall efficiency is through traveling-wave pumping. In this mode, only the localized region through which a wave packet passes is pumped to inversion. This has been demonstrated over a 1-m length with a transversed-discharge pumped H_2 laser[59-61], illustrated in Fig. 23 of this Chapter. Such operation has been demonstrated at wavelengths as short as 1100 Å. This technique has recently been adapted to an obliquely incident laser-driver (see Section 6.1 and Ref. 26 of Chapter 5), and may be applied to axially driven pulsed-power-drivers in the future.

2.4.7. Temperature Effects

The temperatures (electron and ion) in the x-ray laser plasma medium can directly influence the gain achieved. The main positive effect of the electron temperature is to govern the pumping rate. Two opposite cases can be cited as examples. For electron-collisional pumping, the pumping rate increases rapidly with a rising electron temperature (see Section 1 of Chapter 3). On the other hand, for recombination pumping, the rate of pumping varies as T_e^{-2}, i.e., rises rapidly with decreasing temperature in a cooling plasma (Section 1 of Chapter 4).

The electron temperature also enters into the high-density collision limits described in Section 2.4.2a through the electron-collisional mixing rates. For photo-pumping such as described in Chapters 3–5, the electron temperature in the source plasma influences the gain in the lasant plasma indirectly through the brightness temperature.

The ion temperature in the plasma influences the gain achieved according to $T^{-1/2}$ for Doppler broadened lines (Section 2.2.2). Hence, it is desirable to have as low an ion temperature as possible. This may or may not be compatible with the plasma's electron temperature, depending on the particular pumping mechanism and plasma structure. At the high densities needed to achieve significant ASE, the electron and ion temperatures are often similar.

2.5. Pump Power Required

Besides such restrictions as opacity described above, another fundamental limit on the volume of the laser medium is set by the power density required and available. A minimum value for this, which cannot be avoided, may be obtained from that which is lost in spontaneous radiation from the upper laser state, i.e., (for a total upper level decay rate D_u),

$$\frac{W}{V} = N_u D_u h \nu_{uo} \quad \text{Watts/cm}^3. \tag{38}$$

In many current systems the plasma thermal (NkT), dynamic (e.g., expansion), and ionization energy densities will combine to equal or exceed this. Assuming that these can be minimized by careful plasma design, such that the power density available in the excited states of the lasing ion can be optimized, we can derive the following guidelines.

2.5.1. Analysis

We begin by assuming that the (total) decay rate D_u is radiative. It can therefore be approximately replaced in Eq. (38) by $A_{uo} + A_{u\ell} \approx 2A_{u\ell}$. (This is particularly true for the hydrogenic-ion $n = 3$ to $n = 2$ lasing transition used as a previous example.) Replacing $N_u A_{u\ell}$ in Eq. (38) by the gain coefficient G (times the length L of the medium) using Eqs. (8) and (10), this leads to

$$\frac{WL}{V} = 16\pi \frac{(GL)hc^2(\Delta\lambda_{u\ell}/\lambda_{u\ell})}{\lambda_{u\ell}^3 \lambda_{uo}}, \tag{39}$$

where, again, $\lambda_{u\ell}$ is the lasing wavelength.

This power density is strongly dependent on wavelength. For example, within the approximation that $\lambda_{u\ell} \approx 6\lambda_{uo}$, the required power density is proportional to $\lambda_{u\ell}^{-4}$. Assuming as in previous examples that $GL = 5$ and $\Delta\nu_{u\ell}/\nu_{u\ell} = 3 \times 10^{-4}$, Eq. (39) becomes numerically

$$\frac{W}{V/L} = \frac{W}{a} \approx \frac{2 \times 10^{19}}{\lambda_{u\ell}^4} \frac{W}{cm^2}, \tag{40}$$

for the laser wavelength $\lambda_{u\ell}$ expressed in Angstrom units. Here a is the cross-sectional area of the medium. For a short wavelength and limited pumping power, a small area is required. Hence, for $\lambda_{u\ell} = 100$ Å the power flux deposited in the plasma is required to be at least 10^{11} W/cm² (or the same number of Watts per cubic centimeter of power density for a 1 cm length). For a diameter of 100 μm this is equivalent to a total power deposited just in the upper laser state of ≈ 10 MW. With the power scaling inversely as the fourth power of the wavelength, the stored power would have to be increased to 100 GW at a wavelength of 10 Å and for the same diameter. (A rapid rise of power density with wavelength also pertains to the plasma thermal energy NkT.)

2.5.2. Wavelength Dependence

Equation (40) is tabulated in Table 4 for various wavelengths[13], along with an estimate of the minimum power that must be applied. This latter estimate is based on an assumed conversion efficiency of 1%. These quantities are more than sufficient to ionize and heat the medium into a plasma state. Hence the assumption of a plasma medium for x-ray lasers used throughout most of this book is valid.

Therefore, for a given amount of available pumping power and for a certain desired overall gain (determined by the GL product), the lasant

Table 4

Power per Unit Cross Sectional Area in
W/cm^2 versus Wavelength

λ_{ul} [Å] =	1	10	100	1000	3000
Plasma:	10^{19}	10^{15}	10^{11}	10^{7}	10^{5}
Total^a:	10^{21}	10^{17}	10^{13}	10^{9}	10^{7}

^{a} Assuming 1% energy transfer to the plasma.

cross-sectional area "a" must be restricted. It is still possible to trade length L for gain coefficient G, keeping the product constant. This continues to be an important design possibility, i.e., it may permit operation at lower densities and longer lengths for a sufficiently long duration (as discussed in the previous section).

2.5.3. Pumping Efficiency

These pump-power requirements for a chosen gain product GL are in effect an expression of pumping efficiency. The ratio of GL/P from various experiments is plotted versus wavelength for numerous amplifying systems in Fig. 1 of Chapter 7. One experiment[62] clearly stands out, requiring > 100-times less pumping power than other schemes. This is a recombination experiment using a long wavelength, long pulse driver focused axially onto a target in a solenoid. The experiment is described in more detail in Section 2.3 of Chapter 4. Unfortunately the mechanism is not sufficiently well understood to adapt to other devices and drivers[63].

3. ANALYSIS TECHNIQUES FOR X-RAY GAIN IN PLASMAS

X-ray gain experiments involve high-power pumping devices and are very time-consuming and expensive. Hence, it is important to analyze a particular approach as completely as possible before launching a proof-of-principle experiment. Extensive modeling can now be performed with sophisticated numerical codes, as described below.

It can be very expensive and time consuming to develop a code for a particular ion. Therefore, it is important to have rather general analytical tools available for initial proof-of-principle estimates. Such an approach is used in describing various x-ray laser schemes in Chapters 3–5 below. We start here with such a simplified steady-state analysis. We then follow this by describing briefly some more formal approaches that serve as the bases for the powerful plasma-hydrodynamic and atomic-kinetics numerical codes that have been developed. For a detailed description of such codes, the reader is referred to the specific articles on this subject.

3.1. Simple Steady-State Analytical Models

First-approximation steady-state analyses involving only major pumping and decay rates for the upper-laser level, and in some cases for the lower-laser level, are useful for initial evaluations. In the following subsections, an illustration of this is given which is particularly useful in discussing quenching by radiative trapping. It is based on hydrogenic C^{5+} lasing on the $n = 3$ to $n = 2$ Balmer-α line, an example also used extensively in Chapters 3–5. This sample analysis[64] is created around the basic energy levels shown in Fig. 2.

3.1.1. Rate Balancing in Steady State

The primary goal of a x-ray laser gain analysis is to determine (and maximize) the absolute upper-state density N_u, as well as its value relative to that of the lower state, N_1. This density enters first as a direct linear factor in the gain formulae [e.g., Eq. (8) of Section 2.2.2]. It is usually dealt with in a ratio to a ground state density such as N_o (or N_f), e.g., by replacing N_u by $N_o(N_u/N_o)$. This is done because an absolute value of N_o is more often known from measurements or calculations of ionization balance (discussed below). In the simplest case $N_u/N_o = P_{ou}/A_{u\ell}$ for mostly radiative decay to the lower state [see Eq. (19)]. This approximation, sometimes with collisional depopulation $C_{u\ell}$ added, is used throughout the analyses in Chapters 3–5 to estimate N_u.

The upper-state density enters the gain equation again through the ratio N_u/N_ℓ in the inversion factor $F = [1 - g_u N_\ell / g_\ell N_u]$. Here the density ratio is often influenced by collisional-mixing rates that tend to equilibrate the excited levels (Section 2.4.2a). Such mixing can lead to $F \leq 0$ and a quenching of the population-density inversion and gain. Hence, for this dependence, it is usually sufficient to determine the relative value of N_u. This will become clearer from the following simple example, based on a three level system[64].

In steady-state, the rate of population of a particular level equals the rate of depopulation. Referring again to Fig. 2 and ignoring other levels, we can express this as a simple rate equation for the lower-laser level:

$$\frac{dN_\ell}{dt} = N_u[N_e C_{u\ell} + A_{u\ell} + \varepsilon N_e C_{f\ell}] - N_\ell[N_e C_{\ell u} + G_{\ell f}(\tau_c) A_{\ell f}] = 0. \quad (41)$$

Here $\varepsilon = (N_f/N_u)$. The transition probabilities for spontaneous emission are again designated by $A_{u\ell}$ and $A_{\ell f}$. The multiplying factor $G_{\ell f}(\tau_c)$ is the "escape factor" at line center, discussed in Section 2.4.3. It is a function of the opacity of the ℓ to f transition and will be used to demonstrate the effect of radiative trapping of the emission from this transition. The parameters $C_{f\ell}$ and $C_{\ell u}$ represent electron-collisional excitation (ce) (and deexcitation, $C_{\ell f}$, $C_{u\ell}$) rate coefficients, usually expressed in cm^3/sec. These transitions are induced by the inelastic scattering of free electrons. In terms of the cross-section σ_{ce} for such a reaction, $C = \langle \sigma_{ce} v_e \rangle$. Here, the average is usually assumed to be over a Maxwellian distribution of free-electron velocities v_e for an electron temperature T_e in the plasma. The product $N_e C$ represents the corresponding rate. This then leads to an expression for the inversion factor F of

$$F = 1 - \frac{N_\ell}{N_u} \frac{g_u}{g_\ell}$$

$$= 1 - \frac{g_u}{g_\ell} \frac{(N_e C_{u\ell} + A_{u\ell} + \varepsilon N_e C_{f\ell})}{(N_e C_{\ell u} + G(\tau_c) A_{\ell f})}. \qquad (42)$$

3.1.2. Hydrogenic-Ion Example

It is interesting and illustrative to carry this a step further, and to analyze the gain criteria of $F \geq 0$, for varying plasma electron temperatures T_e and densities N_e and for a hydrogenic-ion lasant of nuclear charge Z. As is done extensively in Chapters 3–5, we adopt the convenient Bethe-Born effective Gaunt factor formulism for the electron-collisional excitation rates [Eqs. (2) and (3) of Chapter 3]. We also introduce the reduced variables $N_e' = N_e/Z^7$ and $kT_e' = kT_e/Z^2$. Hence, with the transition probabilities scaling as Z^4, Eq. (42) becomes independent of Z.

3.1.2a. OPTICALLY THIN DEPLETION TRANSITION. The (reduced) electron densities obtained from Eq. (42) versus (reduced) temperature are plotted in Fig. 17 for a "threshold" value of $F = 0$. In this initial example, we assume that $G_{\ell f}(\tau_c) = 1$. This is the case for an optically thin lower-laser-level depletion transition. Two potential lasing transitions are treated, namely, $n = 3$ to $n = 2$ and $n = 4$ to $n = 3$. The curves represent an approximate density boundary between the inversion and non-inversion regions of parameter space. The maximum electron density in the range of $\approx 3 \times 10^{14} Z^7$ cm^{-3} is used as a guide in later analyses (Chapters 3–5).

The gradual increase in density with increasing temperature shown in Fig. 17 is associated with the $T^{-1/2}$ dependence of the collisional excitation

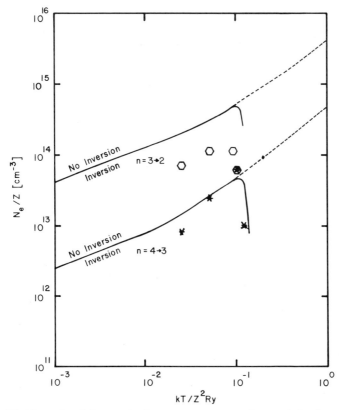

Fig. 17. Limiting reduced electron density versus temperature for population inversion, with no radiative trapping. The o's and *'s represent numerical results from Ref. 65. (From Ref. 64.)

rates. The smooth curve continued dashed is for $\varepsilon = 0$, i.e., an insignificant population in N_f. When ε is increased for typical laser-produced-plasma semi-equilibrium conditions, the curve drops abruptly near $kT' = 0.1$ as shown. This is because of enhanced lower-laser level population from the ground state by electron-collisional excitation.

That such an analysis is reasonably accurate is shown by comparison with some results of a more detailed numerical analysis[65], as indicated in Fig. 17. (In this figure, Ry represents the hydrogenic Rydberg in energy units, equal to 13.6 eV.) Also, the limits have been verified experimentally in laser-produced plasmas[66,67].

3.1.2b. OPTICALLY THICK DEPLETION TRANSITION AND RADIATIVE TRAPPING. The lower-laser-level radiative-trapping problem referred to in Section 2.4.3 of this chapter can be illuminated in this simple analytical model by allowing the escape factor $G_{ff}(\tau_c)$ to become less than unity with increasing opacity. Here again, τ_c is the optical depth at line center given by Eqs. (30) to (31). For hydrogenic ions, with λ_c scaling as Z^{-2} and $\mu \approx 2Z$ and substituting $N_f = N_e(N_f/N_e)$, we can define an optical depth reduced parameter $\Delta = dZ^{4.5}(N_1/N_e)$. This is written for a plasma depth (diameter) "d" and for $f = $ unity, the ground state. A typical value[64] for the density ratio is $N_1/N_e = 1/Z$, so that $d \approx \Delta/Z^{3.5}$. Hence, τ_c [and $G_{lf}(\tau_c)$] can be expressed independently of Z by specifying Δ instead of d.

The effect of the inclusion of opacity is shown in Fig. 18, for an inversion factor of $F = 0.3$. As D increases, inversions are achieved only at lower densities, which implies a need for longer lengths for significant gain. It is seen that serious degradation of inversion conditions occur for $\Delta = 1$ and $\Delta = 100$

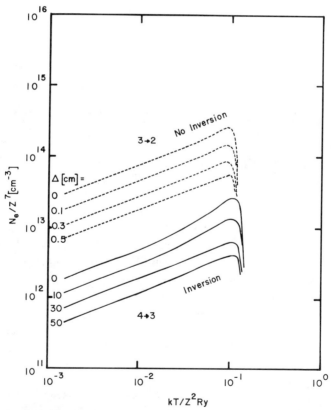

Fig. 18. Limiting reduced electron density versus temperature for population inversion with $F = 0.3$ and radiative trapping included for a reduced depth $\Delta = dZ^{3.5}$. (From Ref. 64.)

for $n = 3-2$ and $n = 4-2$ transitions, respectively. From the above relations, these translate into maximum depths or x-ray laser diameters of 20 μm and 2 mm, respectively for carbon ($Z = 6$). (The first value was found in Section 2.4.3a.)

3.2. Numerical Rate Equation Analyses

Analyses such as are described above and used to illustrate various x-ray laser schemes in later chapters are obviously only first-order approximations and usually tend to be optimistic. They are quite useful for determining the relative merit of a particular scheme. They are also valuable for gaining a physical feeling for the overall feasibility of a particular approach.

To proceed further, full rate equation analyses are required. This is both for ionization balance to determine the density of a particular ionic species and for excited-level population distributions to determine the population-inversion densities and gains. With such a numerical analysis, the time dependence of a non-steady-state gain condition can be modeled also. Collisional rates involved are dependent on the energy of the particles, determined in plasmas by the thermal temperature, as well as the density of perturbing particles. Hence the plasma conditions must also be known.

3.2.1. Hydrodynamic Codes

Such modeling is usually done in two parts. First hydrodynamic modeling of the plasma evolution predicts the dynamics and flow of important plasma parameters such as electron and ion temperatures and densities as functions of time. This has evolved into a very sophisticated field of its own. One of the most well known and extensively used codes developed at Lawrence Livermore and Los Alamos National Laboratories is termed LASNEX. A description of such codes is not appropriate for this book. A reader interested in pursuing this topic for x-ray laser modeling is referred to numerous citations in Ref. 68.

3.2.2. Atomic Physics and Kinetics Codes

As a second step, the atomic and ionic modeling of x-ray laser interest is often performed either with a post-processor to the hydrodynamics codes or simply by assuming a set of plasma parameters (i.e., without using any hydrodynamic code).[68]

3.2.2a. IONIZATION BALANCE.   The densities of atoms in particular stages of ionization are first determined by rate equations for balance between ionization and recombination for, say, an initial ground state of density N^i in an ionization stage $i+$ $(i = 1, 2, 3, \ldots, Z - 1, Z)$. Hence, the $Z+$ ion is completely stripped. Such an equation takes the form

$$\frac{dN^i}{dt} = N_e \left\{ N^{i-1}I^{i-1} + N^{i+1} \left[R_{rr}^{i+1} + N_e R_{cr}'^{i+1} + R_{de}^{i+1} + \left(\frac{N_A}{N_e} \right) R_{ct}^{i+1} \right] \right.$$

$$\left. - N^i \left[R_{rr}^i + R_{cr}^i + R_{de}^i + \left(\frac{N_A}{N_e} \right) R_{ct}^i \right] \right\}, \tag{43}$$

Here I and R represent ionization and recombination rate coefficients. The subscripts rr, cr, de and ct stand for radiative, collisional and dielectronic recombination, and finally charge transfer (from neutrals), respectively. These are further defined in Chapters 4 and 5, where R is replaced by the pumping rate P. The collisional recombination term is primed, where $N_e R_{cr}' = R_{cr}$ is the true rate coefficient. It is written in this way to emphasize the additional dependence on electron density for the three-body process. The densities N_e and N_A are those of the perturbing electrons and neutral atoms, respectively. Again, in steady state, the overall rate on the left side in Eqs. (43) can be set equal to zero. Then, the relative densities of ion species may be determined. Again a major strength of full numerical modeling is the time dependence obtainable with time-varying plasma conditions.

Solution of equations of the type shown in Eq. (43) can produce the relative densities of ion species in a heating and subsequently cooling plasma.

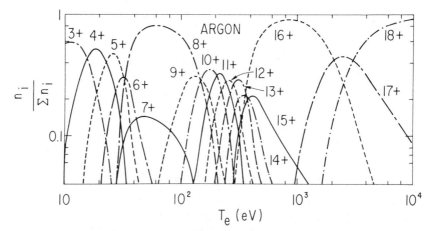

Fig. 19.   Fractional ion abundances calculated for an optically thin argon plasma versus electron kinetic temperature in eV. The particular ionic species are indicated. (From Ref. 70.)

This translates into a time-dependence when the plasma parameters are known as a function of time. An example[70] of the heating phase is shown in Fig. 19 for the various ionic-species in an optically thin argon plasma. This particular example shows the extended temperature range available for the closed-shell Ne-like and He-like (as well as hydrogenic) species. This characteristic can be of significant importance for extended lasing periods (see Sections 1 and 2 of Chapter 3).

3.2.2b. EXCITATION.   The second task is to model the particular kinetics of excitation and deexcitation for specific energy levels in a particular ion. This can be separated from the ionization balance since the rates for excitation and deexcitation are generally much higher than for those between ion stages. A general form for a rate equation for one particular excited level "n" is

$$\frac{dN_n}{dt} = \sum_k \{N_k[(N_e + N_p)C_{kn} + A_{kn}] - N_n[(N_e + N_p)C_{nk}]\}$$

$$+ N_e \left\{ N^{i-1}I_{on}^{i-1} - N_i I_n^i + N^{i+1}\left[R_{rr} + N_e R_{cr}' + R_{de} + \left(\frac{N_A}{N_e}\right) R_{ct} \right]_{on} \right\},$$

$$(44)$$

where $(N_e + N_p)C_{kn}$ and A_{kn} represent respective electron- (plus proton-) collisional deexcitation and spontaneous radiative decay rates from an arbitrary level k to a lower specific level n (and vice-versa).

Proton collisional rates can become important for closely spaced levels if the density of protons is comparable to electrons. Higher charged ions are still less important because of increased Coulomb repulsion with the ion of interest. An example of this is discussed in Section 2 of Chapter 3.

The first line in Eq. (44) includes processes within the particular ionic species of interest. The second line includes the influence on bound level n by adjacent species. This includes collisional ionization out of level n. Also, the radiative transition probabilities in Eq. (44) must be modified for optically thick lines, either by an escape factor $G(\tau_c)$ or by more sophisticated radiative transport modeling. The first option is discussed in Section 2.4.3 of this Chapter and incorporated above. Such escape factors are omitted from Eq. (44) for clarity. With opacity would come photon-induced processes, also excluded from these equations. Innershell processes and associated radiationless Auger or autoionization decay processes could also be included for completeness, if important.

It is not unusual to include hundreds of levels, each with their own rate equation, in a modern kinetics code. This is particularly true for complex Ne-like or Ni-like ions (see Chapter 3, Section 2). Once written, the rate equations are solved simultaneously to get particular energy level populations as a function of time. There are many versions of such codes, using various expressions for the rate coefficients involved. A critical review of these would be extensive and appears to be lacking at this time. For further information the reader is referred to articles that include such modeling[68-71].

3.2.3. Plasma Neutrality

In order to solve these equations numerically it is necessary also to use the plasma neutrality equation

$$N_e = \sum_{k=1}^{i} kN^{k+} \tag{45}$$

for a single-element (i^+)-ionized plasma of ion density N^{i+}

4. PUMPING MODES AND DEVICES

For reasons detailed in Section 2.5, significant x-ray ASE requires large amounts of power concentrated in a line plasma of very small diameter. This has been most successfully accomplished to date using photons as the vehicle for transferring the power from the source into the minute volume. This is largely because of the ease of energy focus of photons, compared to charged particles where Coulomb repulsion limits the energy density achievable. Neutral beams have not yet reached the power levels required.

The technology used for driving x-ray lasers continues to be very much a spinoff from inertial-confinement fusion programs. There energy is also required to be concentrated in short times into very small (spherical) pellets. This is not all a one-way flow of technology. For example, x-ray lasers offer a valuable tool for diagnosing compressed-pellet fusion plasmas at extremely high supra-solid densities (see Section 4.2.3 of Chapter 7).

4.1. Pumping Pulse Length Requirements

For operation in the self-terminating mode, pumping pulse lengths shorter than the decay time are essential. In contrast, for the pure cw mode there is in principle no limit on the length of the pumping pulse. Longer pulses are in fact desirable for operation over long lengths and/or with cavities. However, short pulses are sometimes necessary in the quasi-cw mode. An example is recombination pumping, where rapid cooling following heating is essential, and it is necessary to remove the pumping impulse in a time shorter than the recombination time (see Section 1 of Chapter 4). Generally, shorter pulses require less energy if the pump-power density is held constant (Section 2.5). Again, short pulses are not practical for extended lengths unless a traveling-wave operation is employed.

4.2. Pumping Devices

In this section we describe some popular and successful types of x-ray laser drivers.

4.2.1. High Power Lasers and Focusing Optics

Most x-ray laser schemes require electron densities $\gtrsim 10^{20}$ cm^{-3}, after some expansion. For this, driving lasers of wavelength ≤ 1 μm are most useful. This is because the driving laser couples its energy most efficiently to the plasma through the electrons by the inverse-bremsstrahlung process. The critical wavelength for maximum absorption is given as λ_p in Eqs. (7) and (8) of Section 4.2.3, Chapter 7. There it is shown that $N_e \approx \lambda_p^{-2}$ and $N_e = 10^{21}$ cm^{-3} for a wavelength of 10,600 Å (or 1.6 μm). This is the primary wavelength of high-power Nd:glass lasers. Higher harmonics (0.53, 0.35 μm wavelengths) produce even higher density plasmas, as needed for shorter wavelength x-ray lasers. On the other hand, CO_2 lasers operating at 10.6 μm wavelength couple mainly at a lower electron density, namely 10^{19} cm^{-3}.

4.2.1a. TYPES OF LASERS.   The most widespread success with laser-pumping has been with Nd:glass lasers with pulse widths in the 0.1–2 ns range and with transverse line-focused beams. Energies typically have been in the 0.1–5 kJ range. Specific devices are listed in Table 5. One or more beams are used for providing target irradiances in the 10^{12}–10^{15} W/cm^2 range. A different approach[71] combines a 1 kJ, 75 ns pulse from a 10 μm CO_2 laser with a solenoidal magnetic field (see Section 2 of Chapter 4).

4.2.1b. CYLINDRICAL-LENS FOCUSING OPTICS.   Transverse focusing of the laser-driver beam(s) onto the target has a decided advantage in that both ends of the line-plasma formed are conveniently available for measurement and cavity installation. Examples of this are illustrated[79,92] in Fig. 20, as also outlined on the book cover. Also, as discussed in Section 5.2, the length of the lasing plasma can be varied without altering the incident pumping flux.

Table 5

Typical Characteristics of Some Laser Drivers Used for X-Ray Lasers

Location	Laser Name	Wavelength [μm]	Duration [ns]	Power [TW]	Ref.
CEL[a]	Phebus	1.05, 0.53, 0.35	0.6	6	73
LLE[b]	Omega	0.35	0.6	0.5	74
LLNL[c]	Nova	0.35	0.45	10	54
NRL[d]	Pharos III	1.05	2	0.3	75
Osaka U.[e]	Gekko XII	0.35	0.13	1	76
LULI[f]	Greco IV	1.05	3.5	0.01	77
PPPL[g]		10	75	0.013	72
RAL[h]	Vulcan	0.53	0.07–20	1	63, 78

[a] Centre d'Etudes de Limeil-Valenton, France.
[b] U. Rochester Lab. for Laser Energetics, USA.
[c] Lawrence Livermore National Laboratory, USA.
[d] Naval Research Laboratory, USA.
[e] Osaka University, Japan.
[f] U. Paris Sud & Ecole Polytechnique, Palaiseau, France.
[g] Princeton Plasma Physics Laboratory, USA.
[h] Rutherford Appleton Laboratory, UK.

This permits an accurate determination of a nonlinear response to length variations and the ASE gain coefficient.

Usually a concave cylindrical lens of long focal length and overall size at least that of the laser beam is used to provide a focused line of 100–200 μm in width. A practical advantage here is that the lens system is remote from the target area. The beam is then compressed to the length desired by a spherical focusing lens in tandem with the cylindrical lens. Use of a negative cylindrical lens assures that the line focus falls at the focus of the spherical lens. When properly designed, it is possible to stack a series of cylindrical lenses for additive lengths in discrete amounts. For example, a lenses designed for lengths of 6 and 12 μm can also be used for 18 μm, etc.[50] It is also possible to employ two cylindrical lenses, one of which is rotated to continuously vary the line length[54]. (Naturally, there is an additional loss in driving energy for each lens that is added.)

4.2.1c. CONCAVE-MIRROR FOCUSING OPTICS. An alternate method of focusing into a line of variable length with perhaps less losses is with an off-axis concave mirror[80]. This is illustrated in Fig. 21. Here, a spherical mirror of radius R_m is added to the normal point-focus system, instead of a cylindrical lens. A line image occurs at an off-axis tilt angle θ_m. The length L of the

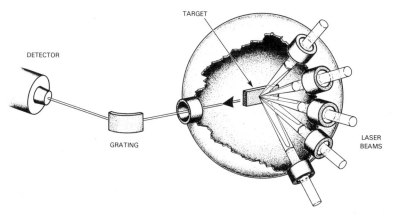

Fig. 20a. Schematic of a x-ray laser pumped transversely by multiple aligned line-focused high power laser beams. (From Ref. 79)

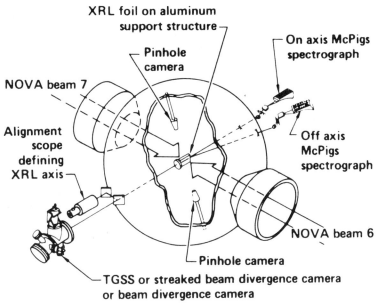

Fig. 20b. Schematic of a x-ray laser pumped transversely by two aligned line-focused high power laser beams from the LLNL Nova facility. (From Ref. 92).

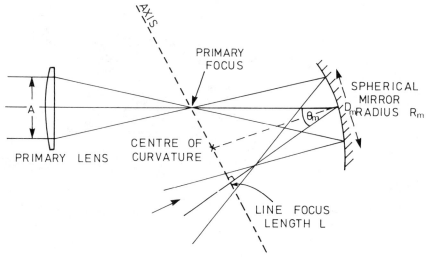

Fig. 21.   Schematic of a line focus design using a concave mirror. (From Ref. 80.)

image is related approximately to the illuminated length D_m of the mirror by

$$\frac{L}{D_m} \approx 1.4 \times 10^{-4} \theta_m^2, \tag{46}$$

for θ_m in degrees. This ratio is unity for $\theta_m = 85°$, so that lengths of many centimeters are possible with available mirrors.

The width w of the image created in this manner using an unaberrated beam is given by

$$w \approx 2.4 \times 10^{-8} \lambda_{u\ell} \frac{R_m}{D_m}, \tag{47}$$

where $\lambda_{u\ell}$ is the lasing wavelength in Å. Both L and w increase with θ_m. Straight lines of aspect ratio $L/w > 10^4$ are predicted to be possible[80].

This system of focusing requires one or more (for multiple beams) mirrors in the proximity of the target. This arrangement can be obstructive to certain transverse diagnostics. This is an additional consideration in planning a total experiment.

4.2.1d. INTENSITY DISTRIBUTION.   The intensity at any point along the line focus is the integral of the input beam along a line normal to the focal line. Therefore, for a circular input beam, the focus, and hence the irradiance, will not be uniform along the line. This is true for either a cylindrical lens or a mirror focusing system. This nonuniformity can be compensated for by shaping the aperture of the incoming beam, rectangularly to first order[80]. Some loss in power delivered to the target is inevitable, however.

4.2.2. Pulsed-Power Drivers

Electrical pulsed-power drivers for x-ray lasers are more direct and potentially more efficient than laser drivers, for larger diameter (millimeter-scale) elongated plasmas. Three types of such drivers, either proposed or currently in use, are described next.

4.2.2a. TRANSVERSE DISCHARGE DEVICES.   Transverse electrical discharges (or electron beams) have been used very successfully for excimer lasers in the near-uv spectral region. By directing the current perpendicular to the axis of the laser, the uniformity of the driving energy deposition as well as the flexibility for extending the length of the lasing medium is preserved. Such devices can also be made to be operated in a traveling-wave mode. This was demonstrated using triggered spark gaps in a flat-plate Blumlein circuit for N_2 and H_2 lasers at vuv wavelengths[59,60] as short as 1160 Å (see Section 2.1.1 of Chapter 3). This technique is illustrated in Fig. 22. Such a

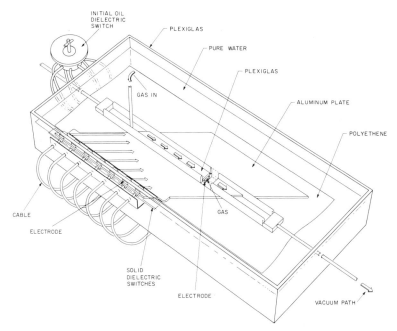

Fig. 22.   Schematic of a traveling-wave excitation system such as used to produce ASE in molecular nitrogen and hydrogen. (From Ref. 60, © 1971 IEEE.)

device is limited in power density in the medium. Hence, it has not generated significant interest for lasing wavelengths in the xuv region.

4.2.2b. LONGITUDINAL Z-PINCH. The axial z-pinch discharge couples more power density from an energy storage facility that does an extended transverse discharge. Hence, higher temperatures, densities and ionization stages are reached in the plasma formed by the axial current. The discharge may take place either through a gas or through a wire(s) or foil. The overall plasma geometry is naturally favorable for x-ray lasers, with plasma lengths measured in centimeters and diameters in millimeters. However, the ends are less accessible for mirrors than with transverse-driven devices.

Pulsed-power drivers used for such z-pinches are considered to be energy-rich sources. They are presently capable of delivering up to 10^{13} W electrical output[81]. Applied voltages and operating currents are in the MV and MA ranges (see Table 6). They operate in a longer pulsed (20–50 ns) mode than do lasers. Hence, the plasmas formed are more suitable for multi-pass cavity operation. Specific device capabilities are listed in Table 6. Such powers are usually larger than those for laser drivers (see Table 5). However, again, the volumes are larger, so that the power density P/V is comparable.

X-ray K-emission of up to 10^{12} W is possible from pulsed-power z-pinch discharges[81]. This implies an overall conversion efficiency as large as 10%. For a typical volume of 0.1 cm^3, this results in $P/V \approx 10^{13}$ W/cm^3. This translates well into the $\lambda \lesssim 100$ Å range for lasing (see Table 4). This assumes that the power can be concentrated in an upper laser state.

Table 6

Typical Characteristics of Some Pulsed Power Drivers[a] Used for
Imploding Z-Pinch X-Ray Lasers

Location	Device Name	Power [TW]	I_{max} [MA]	X-Ray Yield [kJ]	Ref.
MLI[b]	Blackjack V	10	4.6	50	51
NRL[c]	Gamble II	1	1.5	4	82
PI[d]	Double Eagle	8	3	15	83
PI[d]	Pithon	5	3		84
SNL[e]	Proto II	3	3	2.3	85

[a] Adapted in part from Table 1 of Ref. 81.
[b] Maxwell Laboratories, Inc., San Diego, CA.
[c] Naval Research Laboratory, Washington, DC.
[d] Physics International, Inc., San Leandro, CA.
[e] Sandia National Laboratory, Albuquerque, NM.

High current (up to 10 MA) electron beams are also possible. They use such pulsed-power devices for target irradiation in vacuum. This mode of beam-target operation has received limited attention for x-ray lasers, however, compared to the electrical-discharge z-pinch mode.

4.2.2c. CAPILLARY DISCHARGE.   Electrical discharges confined in a channel such as a capillary of 1 mm or less diameter may be applicable for x-ray lasers. The pulse duration in most of such devices tested to date is in the 100s of nanoseconds range. Because of interactions with the walls on such time scales, the temperature has been limited to a few tens of eV. Some promise of a temperature as high as 250 eV is indicated[86] in a short pulse (≈ 20 ns) capillary discharge through teflon. In this device, an x-ray spectrum of hydrogenic fluorine obtained with a crystal spectrograph indicates a possibility of a population inversion between excited states (see Section 5.1), although the film used and its characteristics are unknown. Also, opacity must be taken into account in analyzing the spectrum shown in this reference.

5.   GAIN MEASUREMENT METHODOLOGY

Determination of the presence of significant net gain in a specific experiment is not always a simple matter. In the presence of a high gain coefficient, and/or with an efficient cavity, typical laser beam and spectral qualities such as brightness, collimation, coherence and line narrowing are readily apparent. Usually however, xuv and x-ray gain coefficients are marginal and cavities are inefficient. Hence, a good deal of caution must be exercised in concluding the existence of net gain in a particular experiment. In the following sections various methods along with precautions are described for verifying low-to-moderate values of gain from spectral intensity measurements.

5.1. Population Inversion and Low Gain

As a precursor to amplification demonstrations, population-density inversions are often measured. Relative measurements of the ratio N_u/N_ℓ are the most straightforward. Such measurements (along with known statistical weights) yield directly a value for the population inversion factor F in Eq. (9). Then, an accompanying measured absolute value for the upper state density N_u provides a value for the gain coefficient G, large or small, through Eq. (8). This assumes that the cross-section σ_{stim} given in and following Eq. (10) is known. This cross-section may depend on line width measurements or

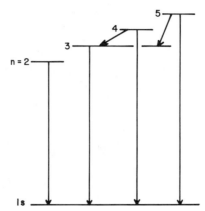

Fig. 23. Energy levels for a hydrogenic ion showing transitions from the excited levels useful for population inversion measurements.

assumptions (e.g., Doppler broadening at a particular ion temperature). This basic spectroscopic procedure provides gain coefficient determinations in tenuous plasma regions, where such resonance lines are optically thin and where the gains are naturally quite small. From there scaling in conjunction with modeling can proceed for advancement to significant amplification experiments at higher densities.

For laser transitions between levels coupled to the ground state (o) through optical transitions, the ratio $I_{uo}/I_{\ell o}$ of emissions from resonance lines provides immediate evidence of population inversions through

$$\frac{N_u}{N_\ell} = \frac{I_{uo}}{I_{\ell o}} \frac{A_{\ell o}}{A_{uo}} \frac{h\nu_{\ell o}}{h\nu_{uo}}, \tag{48}$$

as long as the frequencies (wavelengths) and transition probabilities are known. For resonance lines, the wavelengths are usually close in value, so that variations in instrumental sensitivity are not significant.

An example of this technique is illustrated by the energy-level diagram in Fig. 23 and the spectrum in Fig. 24. This is for the hydrogenic-ion CV and CVI spectra (C^{4+} and C^{5+} ions). Intensity anomalies associated with population density inversions between the $n_u = 4, 5$ and the $n_\ell = 3$ levels are quite noticeable[87]. Here the population inversion was associated with electron capture by stripped C^{6+} ions, forming hydrogenic C^{5+} excited states (see Chapter 4, Sections 6 and 7). The weighted density ratio was deduced to be $N_u g_\ell / N_\ell g_u \approx 3.5$ and the inversion factor $F \approx 0.7$. Absolute measurements extrapolated from the near-ultraviolet region using a set of branching ratios (Section 6.6.2 below) provided a measured[88] value for the gain coefficient of $G \approx 0.01$ cm^{-1}. This was in a low-density region of an expand-

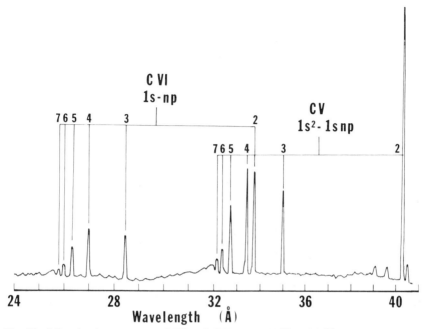

Fig. 24. Microdensitometer scan of CV and CVI spectra (C^{4+} and C^{5+} ions) from a laser-produced plasma, showing intensity anomalies for $n = 4-1$ and $n = 5-1$ spectral lines associated with population inversion. (From Ref. 87.)

ing laser-produced plasma, which then must be extrapolated upwards to significant amplification in regions closer to the target at higher densities. Similar measurements have been performed in other ions[67], and with recombination pumping (see Table 1, Chapter 4).

In some cases of interest, such as the very successful 3p-3s amplification method discussed in Sections 1 and 2 of Chapter 3, the upper-laser-state is metastable to radiative decay. Then spontaneous emission of intensity I_{ul} from the laser transition replaces I_{uo} (along with the associated parameters) in Eq. (48). For calibration, a branching ratio over an extended wavelength interval spanning the 3p-3s xuv laser and 3s-2p lower-level x-ray depletion energies is required. This may be provided *in-situ* in this case by a 3d-3p and 3d-2p branch with similar energies, as shown in Fig. 25.

5.2. Variation of Lasant Length

When significant ASE gain coefficients of $G \gtrsim 1-2$ cm^{-1} are present at high densities, the most straightforward, accurate and convincing evidence of amplification is an observation of exponentiation of intensity with length L of

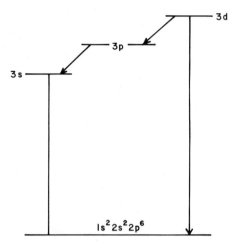

Fig. 25. Energy levels for a Ne-like ion showing transitions from the 3p and 3s excited levels useful for population inversion measurements. Also shown are the 3d-3p and 3d-2p transitions useful for a branching ratio instrument calibration.

the gain medium. The specific response of the intensity to changes in length varies from a simple $\exp(GL)$ relation for the throughput of an external source to the ASE relations defined in Section 2.2.1. A decided advantage of this technique is that only relative-intensity measurements are required from a single axially directed spectrograph.

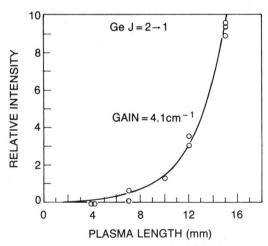

Fig. 26. Relative line intensities vs. plasma length for two 3p-3s lasing lines in Ne-like Ge^{22+} ions. The data points shown are fitted to a theoretical curve to obtain the gain coefficient indicated. The transitions are shown as B and C in Fig. 5 and Table 1 of Chapter 3 (Adapted from Ref. 89.)

This technique was mastered in experiments[55,75] on the Ne-like 3p-3s lasers driven transversely by a high power visible laser (Section 2, Chapter 3). In this form of plasma heating, the length can be conveniently varied without altering the pumping energy distribution along the length of the medium. This is accomplished simply by varying the length of the target. (This is not possible for axial excitation either by a laser or by an electrical discharge.) An example of this technique is illustrated in Fig. 26 for a Ne-like Ge^{22+} lasing ion driven by a Nd:glass laser[89]. The theoretical curve based on Eq. (6) is best-fitted to the data to yield the gain coefficient shown. In this case the 3p-3s lasing lines are essentially non-observable in pure spontaneous emission, so that the theoretical curve can be fitted to the origin at zero length. On the other hand, for lasing transitions such as hydrogenic $n = 3$ to $n = 2$ Balmer-α, used in a number of examples in this book, there is already significant spontaneous emission without gain and such curve fitting becomes more difficult.

5.3. Orthogonal Intensity Measurements

Often it is not possible to vary the length of the x-ray lasing plasma without altering the plasma and excitation conditions. Examples of this are plasmas created by axially directed laser beams or electrical discharges. When this situation arises, an alternative technique for measuring gain is by the ratio of orthogonal axial and transverse intensities[90], i.e., I_a/I_t. This is more difficult to accurately quantify for an absolute gain measurement. It is necessary to know the exact region of view and the collecting optical parameters in each direction. It is also necessary to accurately calibrate the spectrographs relative to each other. For marginal ASE gain coefficients G (~ 1 cm^{-1}), such a calibration must be very reliable. Usually the entire length of the plasma is viewed in the transverse direction. The method consists of measuring an intensity ratio developed using Eq. (3) above, which is

$$\frac{I_a}{I_t} = \frac{\exp(GL - 1)}{GL}. \tag{49}$$

A diagram of this technique[90] is shown in Fig. 27.

5.4. Negative Absorption Measurement

A fourth method of demonstrating and measuring ASE in a plasma is by negative absorption of the emission from one plasma within the gain medium

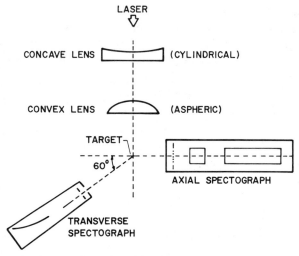

Fig. 27. Schematic of an experiment in which gain was measured simultaneously on- and off-axis. (From Ref. 90.)

consisting of a second plasma[91]. This technique only involves axial measurement with a single spectrograph, as was the case in Section 5.2. However, it does require two independent plasmas. The spectral line on which gain is present then shows up as a reversal on an absorption spectrum. The effect must be distinguished from self-absorption and self-reversal if the laser transition is strong.

5.5. Multipass with Reflectors

Additional techniques become possible when mirrors are available. Two of these will now be described.

5.5.1. Double Pass with One Mirror

A conceptually straightforward method similar to increasing the length is to reflect the output beam from one end of the lasing plasma through 180°. Even if the reflectivity is only $\sim 20\%$, an exponential increase of $\exp(5) = 148$ over a 1 cm return path will compensate for the loss at the mirror. This has been demonstrated in two experiments[62,92], using multilayer mirrors such as described in Section 2.3 of Chapter 1. This technique can be extended to two mirrors which becomes a laser cavity within certain limitations, as discussed in Section 2.3 above, as well as in the following section.

5.5.2. Multipass in a Limited Cavity

As discussed earlier, truly efficient cavity operation in the x-ray region currently is limited by the low reflectivity of the reflectors. Also, the pulse length limits the number of passes of the beam through the cavity. The most efficient soft x-ray mirrors are multilayer dielectrics, which are quite fragile. Hence they must be placed at considerable distance from the expanding lasma for survivability. This further limits the number of passes $\mathcal{N}_c$ of the laser beam through the gain medium, such that

$$\mathcal{N}_c = \frac{ct_g}{z_m}. \tag{50}$$

Here t_g represents the duration of the gain and z_m is the separation of the mirrors. Typical values are $t_g = 1$ ns and $z_m = 10$ cm, resulting in $\mathcal{N}_c = 3$ (or ~ 0.3 ns/pass through the cavity). Hence, steady-state cavity operation, where $\mathcal{N}_c \rightarrow \infty$ and modes are fully developed, is currently not feasible.

When one is attempting to verify low levels of gain with a limited cavity, there is an added advantage to having the length of the cavity considerably greater than the length of the gain medium. Amplified emission from each additional pass through the medium is delayed by an interval $z_m/c \approx 0.3$ ns in the example above. With time resolution at least as accurate as this interval, one can determine the gain on each pass and hence the progressive efficiency of the cavity and the medium. This was demonstrated in a recent experiment performed near 200 Å, where gain over 3 passes was observed in individual pulses[25]. In this experiment, the "rear" mirror had a reflectivity of 20%. The "front" mirror (beamsplitter) was formed on a 400-Å-thick silicon nitride substrate and had a measured reflectivity and transmission of 15% and 5%, respectively.

Often such temporal resolution of individual passages of the beam through the gain medium are not feasible. Without adequate time resolution, the output following individual reflections are indistinguishable and thereby accumulate. A typical example of this is an early gain experiment at 600 Å in calcium, which incorporated two gold-coated mirrors and a grating to split off a portion of the beam[26].

In performing time-integrated measurements of the overall enhancement, the output with a fully aligned cavity is typically much larger than that in which the cavity is intentionally misaligned (e.g., the rear mirror obstructed). When such a comparison is used and the amplification per pass is low, certain additional precautions should be taken to distinguish an increase in output from accumulated gain to any from inherent effects due strictly to the cavity itself[93]. Such cavity enhancement is simply a summation over $\mathcal{N}_c$ passes of pulses exiting the cavity through a beam splitter of low transmission. A simple tracing of individual rays shows that the full cavity enhancement Ψ_{cav} compared to the case where the rear reflections are eliminated (by, e.g., obstructing the rear mirror) is given by

$$\Psi_{cav} = \sum_{j=0}^{\mathcal{N}_c} [R_= \exp(GL)]^j. \tag{51}$$

This is for a net gain product of GL and equal reflectances $R_=$ ($=R_1 = R_2$) for both mirrors. The result is not significantly difference for different front and rear R-values, as long as they are approximately equal[93].

For x-ray lasers with finite $\mathcal{N}_c$, this cavity enhancement can be approximated by a partial sum. At threshold (i.e., for $GL = 0$), this becomes

$$E_{cav} = \frac{(1 - R^{\mathcal{N}_c+1})}{(1 - R)}. \tag{52}$$

In the unlikely event that $\mathcal{N}_c$ becomes very large and approaches infinity, Ψ_{cav} becomes[93] $1/(1 - R)$. In fact, for $R < 0.4$, Eq. (52) is close to this limiting value as illustrated in Fig. 28. As seen there, when one attempts to identify gain by comparing full-cavity and front-mirror-only cavity signals, signifi-

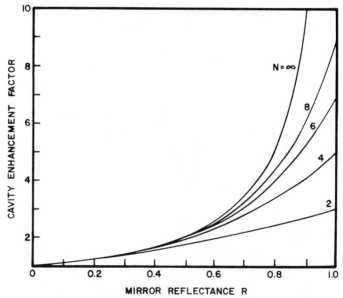

Fig. 28. Cavity enhancement factor versus mirror reflectance for various numbers N of passes of the laser beam through the gain medium.

cant threshold cavity enhancements by approximately factors of two can be expected, even with no gain.

6. MEASUREMENT INSTRUMENTATION

Measurement of gain and power output may involve relative intensities for population inversions, or absolute intensities for power radiated and population densities. X-ray diagnostics of the plasma are also vital in determining the environmental conditions conductive to efficient pumping. In any case, the instrumentation required is of vuv optical and spectroscopic nature.

6.1. Vacuum-UV Imaging Devices

It is important to be able to image the x-rays emitted from the lasing plasma, both for beam concentration and for auxiliary diagnostics. For wavelengths shorter than 1200 Å, conventional lenses are not available for imaging x-ray emitting extended sources. Some alternatives are now described. Further details of x-ray and gamma-ray imaging are covered extensively in the conference proceedings listed below as Ref. 94.

6.1.1. Pinholes and Coded Masks

For imaging an extended soft x-ray source, the simplest method is the pin-hole, camera, illustrated in the top portion of Fig. 29. A photograph of an x-ray laser plasma obtained though a pinhole is shown in Fig. 15. Pinholes 50 μm in diameter formed in metallic foils are typically used. Particular wavelength bands of interest may be isolated, using thin metallic semi-transparent filters. The solid angle of collection is very small. Hence a bright source and/or a sensitive detection system are needed. The resolution improves with decreasing hole diameter; however, the aperture also decreases as the area of the hole. Thus, there is a trade off between resolution and x-ray throughput. There are more exotic focusing systems with larger apertures described below. These are based on Fresnel refraction and grazing incidence reflections of soft x-rays from surfaces. However, the pinhole camera may be the only practical device for hard x-rays and for gamma rays at photon energies greater than 10 keV[95].

A variant on the pinhole camera that has many such apertures arranged in intentionally irregular patterns, with enhanced photon collection, is the coded-aperture mask[95]. It is illustrated in the lower portion of Fig. 29. To understand the use of such a mask, one can consider the extended x-ray

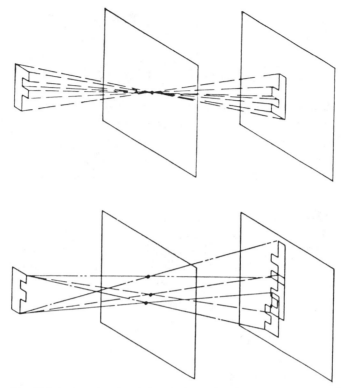

Fig. 29.   Pinhole (upper) and coded-mask (lower) imaging systems. (From Ref. 95.)

source of unknown shape to be a collection of point sources. Each of these casts a different shadow of the intervening mask on the detector. This shadow is known for a particular mask design. By comparing the shadows produced by all possible combinations of point sources with the image recorded, it is possible to reconstruct the image of the source. The key then is to design an optimum mask for reconstruction of the source form. Coded masks with more than 32,000 holes have been used.

This procedure is far from trivial. Fourier and Hadamard mathematical transforms shorten the required computing time significantly and make this technique feasible. Presently, the resolution for soft x-rays is still inferior to the grazing-incidence devices described below, but is improving.

6.1.2. Fresnel Zone Plates

A Fresnel lens or zone plate is a special case of diffraction optics. It is similar to a transmission grating, but with radially increasing line density. This can

Fig. 30.   Free-standing Fresnel zone plate lensing element. (From Ref. 98.)

be seen in the photograph in Fig. 30 of a zone plate to be described below. As with transmission gratings (described below), it exhibits zero, first, and higher orders. Such a zone plate has been found to diffract 25%, 10% and 15% of the incident x-radiation into each of these orders, respectively[96] (the remaining 50% is absorbed in the material). For high zone numbers it also obeys the thin-lens laws of physics. Where transmitting substrates exist, the circular grid is deposited on the surface. However, in the xuv and soft x-ray regions, no such substrate is available. The grid is virtually free-standing with open areas between heavy-metal (e.g., gold) rings.

Fresnel diffraction focuses rays from a point source with a focal length f.l. for first-order diffraction which can be approximated by[96,97]

$$\text{f.l.} = \frac{r_{\mathcal{N}_z}^2}{\mathcal{N}_z \lambda}, \tag{53}$$

where r_n is the radius of the nth ring. The pattern is periodic in r_n^2 and has equal-area annular zones. The outermost zone width is given approximately by

$$\delta r = \frac{r_{\mathcal{N}_z}}{2\mathcal{N}_z}, \tag{54}$$

where $\mathcal{N}_z$ is the total number of zones. Because of the strong (λ^{-1}) chromatic aberration expressed in Eq. (53), the illumination must be narrow-band. It is therefore generally required that

$$\frac{\lambda}{\Delta\lambda} > \mathcal{N}_z, \tag{55}$$

to minimize the effect on the resolution.

From Eqs. (53) and (54), the f-number ($f^{\#} \equiv$ focal length/diameter) is given simply by

$$f^{\#} \equiv \frac{\delta r_{\mathcal{N}_z}}{\lambda}, \tag{56}$$

Hence it depends only on the minimum zone width and the wavelength.

A typical free-standing zone plate focusing element[97,98] such as that shown in Fig. 30 has $\mathcal{N}_z = 500$ zones, with an overall diameter of $2r_{\mathcal{N}_z} = 0.63\ \mu m$ and a minimum zone width from Eq. (54) of 3200 Å. According to Eq. (53), this corresponds to a focal length of f.l. = 4 cm at a wavelength of 50 Å and f-number of $f^{\#} = 64$.

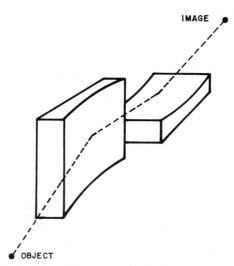

Fig. 31.   Schematic of a Kirkpatrick-Baez grazing-incidence microscope. (From Ref. 102.)

In the past, such microscopic soft x-ray zone plates have found use mainly for high-resolution x-ray microscopes. Their use in x-ray spectroscopy and x-ray lasers is rapidly expanding[96,97.] They are fabricated by various techniques, such as holography, x-ray interference and lithography, evaporation and electron-beam lithography[96,97,99]. The pattern for the zone plate shown in Fig. 31 was created by the last of these techniques, directly in this gold foil. This pattern then was used as a mask for x-ray lithographic replication in PMMA photoresist. This resist pattern in turn served as a mold for production of the free-standing gold zone plate shown. This technique has been used[96,97,100] to generate patterns with linewidths < 1000 Å.

6.1.3. Grazing-Incidence Imaging

X-rays can be focused with a curved mirror by means of total reflection of rays incidence near the grazing angle. Work on x-ray microscopic lenses has been going on since 1929 using such techniques. The most serious problems have been with surface smoothness required and with the astigmatism associated with each reflection. There are several excellent papers on this topic in a 1980 conference proceedings[101].

6.1.3a. KIRKPATRICK-BAEZ MICROSCOPE. Kirkpatrick and Baez[102] in 1948 essentially solved the astigmatism problem by crossing two cylindrical mirrors to obtain two-dimensional focusing. This design is illustrated in Fig. 31. The surface problem has also been essentially solved. Nearly perfect surface finishes better than 10 Å rms are routinely produced now on such materials as silica, virtrious carbon, electroless nickel, silicon crystal, etc.[103] Kirkpatrick-Baez microscopes typically have magnifications ranging from three to eight times and have been operated in photon energy channels ranging from 200 eV to 4.5 keV (wavelengths of 2.8 to 62 Å).

The main application for such a device has been in microscopes, where resolutions of ~ 1 μm have been obtained. They have also found extensive use in recent years for x-ray diagnostics of small laser-fusion pellet plasmas[103]. A major disadvantage is the small solid angle of acceptance ($< 10^{-6}$ steradians for fusion plasma diagnostics). This may be acceptable for x-ray laser use, where present collection solid angles are typically $\sim 10^{-4}$ steradians. The addition of multilayer mirrors (discussed in Section 2.3.2 of Chapter 1) should enhance the present Kirkpatrick-Baez designs[103].

6.1.3b. WOLTER MICROSCOPE. Larger collection solid angles are possible with the Wolter magnifier[103]. This device makes use of axisymmetric mirrors

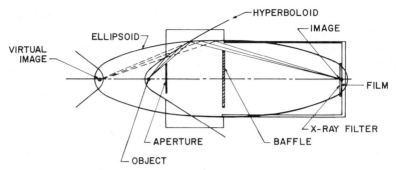

Fig. 32. Schematic of a Wolter hyperboloidal-ellipsoidal axisymmetric x-ray microscope. (From Ref. 103.)

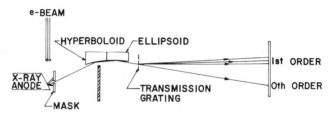

Fig. 33. Schematic diagram of a transmitting-grating spectrograph coupled with a Wolter imaging microscope. The latter focuses the x-rays onto the film. (From Ref. 108.)

made up of conic sections of revolution. In the design shown in Figs. 32 and 33, a hyperpoloid and an ellipsoid of revolution are combined such that they have a common focus. An object placed at the second focus of the hyperboloid is magnified and focused at the secondary ellipsoidal focal point. Practical devices tend to be very long and slender. The requirements for surface smoothness are similar to those mentioned above for the Kirkpatrick-Baez system, but over much larger areas. Magnifications up to 22 times and focal resolutions of 4–5μm have been achieved[103]. Wolter microscopes are known to be difficult to fabricate and very expensive to construct with precision, compared with pinhole/mask apertures and Kirkpatrick-Baez microscopes[104].

6.2. Xuv Grating Spectrographs

6.2.1. Normal Incidence

6.2.1a. REFLECTING GRATINGS.   Spectral dispersion of radiation between approximately 300 and 1000 Å is accomplished with ruled diffraction gratings and with the incident beam at near normal incidence to the surface. In order to avoid additional losses, usually the grating is the single optical element in the spectrograph. It is therefore spherically concave for imaging the entrance slit onto the focal curve. Various spectrographic designs are discussed in detail in Ref. 105. The characteristics of ruled concave gratings are also described in this reference, and more recently in Ref. 106. The grooves of such gratings are usually intentionally cut at an angle to generate a blaze condition. This provides a preferential reflection of radiation into a particular order and wavelength band. Besides improving the overall efficiency, blaze often acts to reduce undesirable radiation from other orders.

More recently, holographic have become available as a substitute for diamond-ruled gratings[107]. More perfect patterns are possible, resulting in less scattered radiation and aberrations. In these gratings, the grooves are usually of a sinusoidal or at least symmetrical shape. Hence blazing is not usually accomplished. There are now some techniques for manufacturing holographic gratings with shaped grooves for achieving some blaze effect. However, the grooves still are not truly triangular as desired.

The xuv throughput efficiency of normal-incidence gratings depends on the rulings and the blaze, as well as the reflectance of the surface. The latter is similar to that for mirrors, multilayer coatings and LSMs discussed in Section 2.3 of Chapter 1. However, great care must be taken in applying coatings to the ruled area, particularly for replicated gratings[107].

6.2.1b. TRANSMISSION GRATINGS. Transmission gratings[108] consisting of fine-mesh lines engraved in thin films offer high aperture and ease of alignment, compared with gratings used at grazing incidence (see Fig. 33). However, besides the small size and aperture at present, the spectral resolution available is limited by the narrow spacing possible. Currently the resolution is approximately 2 Å in the soft x-ray spectral region. This can be compared to a typical resolution of 0.01 Å for a good grazing incidence spectrograph. However, high resolution is often not imperative. Also, x-ray laser lines can be very intense. Hence, transmission gratings can be quite useful in x-ray laser research.

6.2.2. Grazing Incidence

For wavelengths in the shorter 5–300 Å region, dispersion is still usually obtained with concave gratings. In this arrangement, the grating is oriented such that the incident beam enters at a grazing angle of a few degrees to the ruled surface. For angles less than critical for the surface-to-air interface the

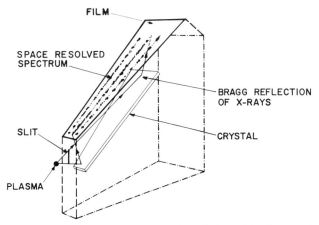

FILM

SPACE RESOLVED
SPECTRUM

BRAGG REFLECTION
OF X-RAYS

SLIT

CRYSTAL

PLASMA

Fig. 34. Flat crystal dispersion of x-rays for spatial-resolution, with a laser-produced plasma as a point source. (From Ref. 119.)

reflectivity can be quite high for wavelengths greater than that corresponding to the particular grazing angle chosen. Such instruments suffer from extreme astigmatism and are difficult to adjust, align and operate. As such, they may eventually be replaced by synthetic microstructures (see Section 2.3.2 of Chapter 1), at least for the longer wavelengths. Astigmatism is sometimes compensated for over a particular spectral band with an auxiliary cylindrical or toroidal condensing lens placed in front of the entrance aperture. A full discussion of grazing incidence soft x-ray spectrographs is available in Ref. 105.

It is often important to spatially resolve the spectral emission from the source, e.g., to ascertain a particular region of emission. This is most readily accomplished over a broad wavelength band with an additional aperture such as an orthogonal slit, oriented perpendicular to the dispersing entrance slit. This auxiliary slit serves the same function as a pinhole camera (discussed above) in one direction, namely along the spectral line. This arrangement is illustrated for a crystal spectrograph in Fig. 34, to be described below. The disadvantage of this spatial-resolving technique is a large loss of sensitivity of the instrument, because of the severe restriction of the collecting solid angle. This may not be a problem when high gain is present and hence strong lasing lines. However, it can be a severe limitation in studies of population densities and inversions with spontaneous emission. This sensitivity limitation can be overcome with a focusing collecting lens as discussed above for a particular band. Hence there is a trade-off between sensitivity and bandwidth in the detection system.

6.2.3. Varied Line-Space Gratings

The traditional ruled grating for instruments described above consists of regularly spaced grooves on a flat or usually concave surface. These are either ruled mechanically or made holographically.

A relatively recent refinement in grating technology is the variable line-spacing grating, in which the spacings between grooves varies smoothly across the ruled area. A review of the evolution of this concept extending back to 1875 is available[109]. Significant technological advances leading to large-scale practical use of such gratings have occurred over the last decade[110–112]. Using a numerically controlled ruling engine, it is possible to rule aberration-corrected concave gratings. Such gratings have been used[110] in the vacuum-uv spectral region in a Seya-Namioka spectrograph. This particular design normally has severe coma-type aberrations at large angles with respect to the grating normal. Designs have been extended to the grazing-incidence region, where aberrations have been even more severe.

In such a grazing incidence design, the spacings between grooves are determined so as to produce minimum defocusing at a fixed detection position as the grating is rotated about its central groove to scan wavelengths. With such a grating it is possible to design[111,113] a spectrograph such that the entrance slit remains in a fixed position as well. This is a very desirable feature for fixed-position sources. A further feature is that the focal plane is approximately flat and perpendicular to the dispersed rays. This also is a distinct advantage for mating with a flat-faced detector such as the cathode of a x-ray streak camera. Such a design can also be reconfigured as a monochromator.

The performance of a grazing incidence spectrograph with a flat varied line-space grating and a collecting mirror recently has been verified experimentally[114]. A resolution of $\lambda/\Delta\lambda = 35,000$ has been measured at a wavelength of 160 Å. This is equivalent to 5 mÅ, and represents an approximately 10 times improvement over the resolution previously obtained with a conventional grazing-incidence spectrograph. Hence, it should be very useful in resolving the spectral line width of the x-ray laser lines described in Chapter 3, Section 2, for example.

6.3. Crystal Spectrographs

Bragg-crystal spectroscopy[115] is a dispersion technique normally associated with the hard x-ray region. However, it can be used with good resolution for soft x-rays of wavelength as long as 25 Å, and with somewhat reduced resolution up to 100 Å with soap-film crystals[116]. Even longer wavelengths are possible[117,118], also with higher reflectivity, using the newer layered synthetic microstructures described in Section 2.3.2 of Chapter 1.

Efficient dispersion takes place at the incident Bragg angle θ_B in the crystal for a particular wavelength λ and dispersive order n_B according to:

$$n_B\lambda = 2d_B \sin \theta_B. \tag{57}$$

This technique is useful for wavelengths as long as $2d_B$, where d_B is the lattice spacing in the crystal. Spectrographs are designed with flat and curved crystals, and with entrance slits or simply with point sources of emission. The latter is far more sensitive than with an entrance slit, considering the lack of available focusing optics. A curved crystal is often used to achieve a variable angle θ_B across the crystal surface for extended spectra. Alignment is not critical, in contrast to grazing incidence spectrographs. In this case, the dispersed beam exits the crystal at an angle of $2\theta_B$. Hence, scattered stray radiation may be specularly reflected into the detector, and often auxiliary filtering is necessary. Spatial resolution can be achieved with a perpendicular slit, as illustrated[119] in Fig. 34, mentioned above.

6.4. X-Ray and Xuv Detection

6.4.1. Passive Detection (e.g., Film, Photoresists)

6.4.1a. PHOTOGRAPHIC FILM.   Sensitive gelatin-free photographic emulsions are available for the vacuum-uv through soft x-ray spectral regions. They are very useful for recording images and large quantities of spectral data. Density versus exposure data[120] for the popular Schumann-Type Kodak 101 emulsion are shown in Fig. 35 for a photon energy of 525 eV (24 Å wavelength). Such films (or plates) are delicate to handle and the range is limited. Saturation occurs at a density of ~ 2, as shown.

A new, more rugged and more sensitive emulsion with a range extending to a density of $\gtrsim 3$ is now becoming available. This film is based on the very sensitive (ASA 100 to 3200) tabular-grain Kodak T-MAX professional film, with the gelatin overcoat omitted.

6.4.1b. PHOTORESISTS.   An alternative to photographic film is a photoresist such as polymethylmethacrylate ("PMMA")[121]. This is an organic polymer that requires an exposure of ~ 1 J/cm^2 in order to break the molecular chain. Such an intensity is $\sim 10^8$ times higher than that required to obtain a photographic density of unity with an xuv emulsion (see Fig. 35). This sensitivity

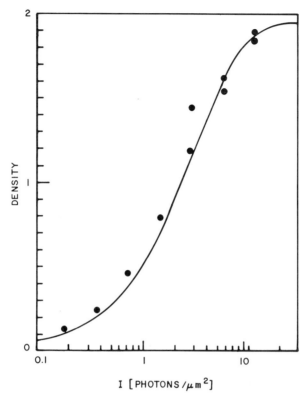

Fig. 35. Measured density vs. exposure data (points) for Kodak type 101-07 film at 525 eV photon energy. The smooth curve was calculated. (From Ref. 120.)

disadvantage is offset by a spatial resolution of ~ 50 Å for PMMA compared to ~ 10 μm (10^5 Å) for Kodak type 101 photographic emulsion. Hence, for the extremely high resolution desired for lithography and biological microscopy (see Chapter 7) photoresists are required, and the high intensities promised by x-ray lasers are important.

6.4.2. Active Detectors

Active electronic photodetectors such as photomultipliers and microchannel arrays, photodiodes, ionization chambers, CCDs and semi-conductors are all responsive to x-radiation to varying degrees. Some have amplification factors as large as $\sim 10^6$. They respond either to x-rays directly or from a fluorescent scintillator. In some cases image recording is totally sacrificed to achieve short and continual temporal resolution. These are discussed further in the following section.

6.5. Temporal Resolution

Often in x-ray laser research it is important to have temporal resolution. This is needed in order to determine the optimum instant and duration of lasing during the pumping period, and hence the peak power. A time-resolving detector can be added to the dispersing instruments described above at the exit slit. In this manner, it becomes possible to isolate and measure the intensity of a specific spectral region as a function of time.

6.5.1. Film with Shutter

When time resolution of < 1 μs is not required, photographic films operating with fast shutters are very useful for collecting large amounts of photographic or spectral data. Such shutters must, however, be synchronized with the pulsed event under study.

6.5.2. Scintillators

For time longer than 1 ns, scintillators coupled with the active detectors described above are very convenient. They are particularly useful whenever the evacuated pressure in the instrument is not sufficiently low to operate a high-voltage active detector without a sealed envelope. Hence, the scintillator can be used over the vacuum enclosure to convert the high energy x-ray photons to visible or near-uv photons transmitted by the window. Various popular xuv scintillators and their time responses and relative sensitivities are compared in Table 7.

The active detectors described above and often used with scintillators have an inherent time resolution as short as a few-tenths of a nanosecond. This is considerably shorter than scintillators. Hence, they are more than adequate when coupled with scintillators. They also can provide improved temporal resolution when used in a direct response mode.

6.5.3. Microchannel Plates (MCPs)

6.5.3a. THE DEVICE. A microchannel plate (MCP), also called a multi-channel electron multiplier array (CEMA), is an array of miniature photo-multipliers[123]. Each element of the array is a channel of typically 8–25 μm diameter formed in a glass fiber. Separations between such bundled capillaries are kept as small as 15 μm. A typical length-to-diameter ratio varies

Table 7

Some Measured Properties of Vacuum-uv Scintillators[a]

Material	Fluor. Pk.[b]	Sensitivity[c]	Decay Time[d]
Thin Films			
Sodium Salicylate	4800	1	7–12
Liumogen	5280	1.1	2.5
Coronene	5025	0.11	3.7–10
p-Terphenyl	3910	0.76	3.7–5.5
Crystals			
Anthracene	4480	1.45	12–20
Stilbene	3840	0.58	3–7
Plastics			
Pilot B	4070	1.45	1.5–2.1
NE 102	4200	0.94	2–2.2
NE 103	4850	0.36	4–11
NE 111	3750	1.02	1.5–2.2

[a] Adapted from Ref. 122.
[b] In Angstrom units.
[c] Relative to sodium salicylate.
[d] Measured in nanoseconds.

from 40 to 100. Such a device is illustrated in Fig. 36. While usually of a flat pancake design, MCPs have also been constructed in "slumped" spherical and cylindrical shapes. The latter lends itself particularly to a spectroscopic focal curve.

The walls of the channels are coated with a semiconductor to optimize the secondary emission and to allow for charge repenishment from an applied

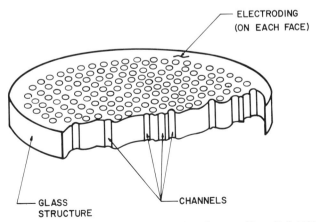

Fig. 36. Schematic of a microchannel plate detector. (From Ref. 123.)

potential source. This coating acts as a resistor chain in a photomultiplier, with a total resistance of $\sim 10^9$ ohms. At the same time, the coated surface serves as a continuous dynode structure. Conducting coatings on each surface serve as input and output electrodes. A typical potential of 1000 Volts is applied, with the more positive potential at the output face (anode). Sometimes, the input electrode is applied in strips which are individually gated to provide several channels of data at different times. Sub-nanosecond gating has been used in spectroscopic analysis of x-ray laser emission[123].

Such devices respond well to direct front-surface charged-particle bombardment, as well as to energetic photons (wavelengths shorter than 2000 Å). They are also used extensively for detecting visible photons when a photocathode is applied. Quantum efficiencies for the MCP surface coatings themselves have been measured in the 10–20% range for soft x-rays[124]. Overcoatings of various materials such as magnesium fluoride have been found to further enhance the xuv photoelectron quantum efficiency at the front surface[125,126]. The relative sensitivity to x-rays with photon energies in the 600 eV to 20 keV range also has been measured recently[127]. Funneling of the input ends of the channels also increases the collection efficiency, particularly at grazing incidence.

Saturated gains exceeding 10^7 are achieved in Chevron configurations illustrated[123] in Fig. 37. Such slanted or even curved designs have been found to suppress ion-drift regenerative feedback from the anode region to the cathode. This is a particular problem when operating at high gain and in base pressures higher than 10^{-6} Torr.

The spatial resolution obtainable with such an MCP array is limited by the size and spacing of the channels and the spread of the output beam of electrons. The electrons emitted are either converted to visible photons and

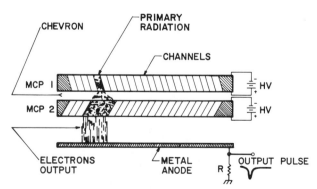

Fig. 37.   Schematic showing operation of a Chevron design microchannel plate. (From Ref. 123.)

then (usually) recorded photographically, or read directly by charge-sensitive devices such as resistive anode encoders. Such detection will be discussed next.

6.5.3b. READOUT SYSTEMS. In a typical readout configuration, the MCP is followed by a phosphor screen to convert the electrons to visible photons. This phospher is usually deposited onto a fiber-optic bundle. This bundle can be closely coupled to a variety of image-processing devices[128]. The simplest and most commonly used is photographic film. Recording on continuous photoresists with no grain structure is also possible. In addition, reticon display can be used, with optical multichannel array (OMA) detection. Furthermore, p-i-n diode and charge-coupled device (CCD) solid-state arrays are available for photon detection and digitizing. None of these readout systems need be totally passive, because some limited "snapshot" time resolution is available by gating of the MCP voltage, as described above.

Continuous temporal resolution can also be added to a dc-operated MCP by coupling a streak camera to the output phosphor-coated fiber optic array. In such an arrangement, the MCP serves both as a photon pre-amplifier and a xuv-to-visible wavelength convertor for a conventional streak camera. Such cameras are discussed in the following section.

6.5.4. X-Ray Streak Cameras

The limitation of MCPs to gated rather than continual time response can be overcome with somethat limited spatial (or spectral) resolution using x-ray streak cameras[119] (see Fig. 38). Picosecond time resolution is possible. Such

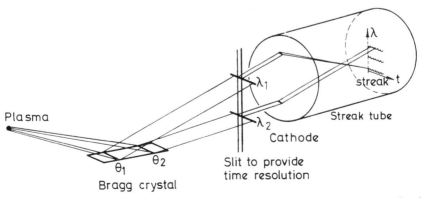

Fig. 38. Schematic diagram of apparatus for x-ray streak spectroscopy of a laser-produced plasma. (From Ref. 119.)

cameras operate on the basis of a swept or deflected electron beam imaged onto a phosphor. Intensifiers are also included. As such, an x-ray streak camera can provide a continuous temporal history of the emission entering through a slit. They often are used with transmitting-grating dispersers of limited spectral resolution but with high aperture and relative ease of alignment (compared to grazing-incidence grating spectrographs). Both thin, transmitting photocathodes as well as front-surface cathodes have been used.

6.6. X-Ray Intensity Calibration

It is important to have xuv detectors that are calibrated on an absolute intensity scale, in order to measure the output power from the x-ray laser. This is no simple task.

6.6.1. Primary Standard

It is preferential to calibrate an entire spectrographic detection system with a standard source. For this, synchrotrons can be used where available to provide absolute calibrations throughout most of the xuv region[129].

6.6.2. Secondary Standard

6.6.2a. STANDARD DETECTOR.   The next best approach is to use a standard detector suitable for monitoring a calibration source essentially in parallel with the irradiation of the x-ray laser detection system by the same source. In that way, the detector records the photon flux entering the detection system over a particular time interval. One example of such a detector which is useful in the $\lambda \leq 100$ Å region is the flow-proportional counter[130] (see Fig. 39). This device delivers a dc photon-count with an output voltage which is proportional to the photon energy (for spectral resolution). This can monitor a stable dc x-ray tube which is also viewed by the x-ray laser detection system.

6.6.2b. BRANCHING RATIO EXTRAPOLATION.   Another example is the branching ratio technique. In this method, two spectral lines from transitions originating on the same upper level u but significantly different frequencies v are compared. One is monitored on an absolute intensity scale at long wavelengths and the other in the xuv region. Often two spectrographs are used.

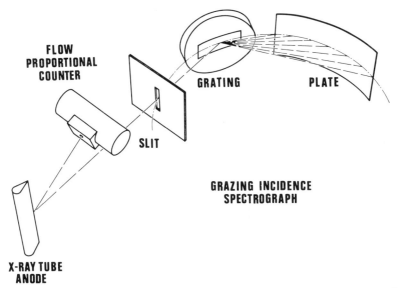

Fig. 39. Schematic of a calibration system (grazing incidence spectrograph) based upon a flow proportional counter as an absolute detector. (From Ref. 130.)

The intensity I of each spectral line is given by:

$$I = N_u A h v, \qquad (58)$$

where N_u is the density in the common upper state. The ratio of intensities is proportional to the frequency v and the transition probability A. If the latter is known, a calibrated emission from the long wavelength line can be transferred to the shorter one by measuring each, providing neither line is optically thick (not an uncommon problem). An alternate description of this method is that the calibrated spectrograph measures the absolute upper state density N_u. From this, the xuv-transition line intensity is obtain from Eq. (58). An example of such branches might be the Balmer- or higher-series along with Lyman-series lines of hydrogen and hydrogenic ions. A compilation of some useful branches is currently in preparation[131].

6.6.3. Standard Detector and Instrument Transmission

A third method is to use a spectrally unresolved but calibrated detector at the exit end of a dispersing system, and then calibrate separately the transmitting efficiency of the system.[132] From the detector viewpoint, the absolute sensitivity of popular xuv and soft x-ray photographic films is quite well established, with limited accuracy as far as reproducibility between samples. Ionization gauges can be operated as absolute detectors for the xuv region and used as standard detectors[105]. Also, the quantum efficiency of a sodium salicylate scintillator is practically constant with wavelength[105]. This phosphor can serve as a calibrated-source extender from the visible, uv or near-vuv regions (see also Table 7). Then it is necessary to determine the instrument transmission, and this can be done by comparing a split direct- and transmitted-beam[130].

6.6.4. Precision Expected

Generally, the overall accuracies of such xuv calibration techniques are on the order of 20% for relative intensities and a factor of two for absolute intensities, unless rather heroic pains are taken. Possible comparisons of various calibration methods in the same wavelength region would be very useful to improve the confidence. However, the overlaps are quite limited, and such data are sparse at present.

REFERENCES

1. L. Allen and G. I. Peters, *Phys. Letters* **31A**, 95 (1970).
2. R. H. Dicke, *Phys. Rev.* **93**, 99 (1954).
3. A. E. Siegman, "Lasers" (University Science Books, Mill Valley, California 1986).
4. G. I. Peters and L. Allen, *J. Phys. A* **4**, 238 (1971).
5. L. Allen and G. I. Peters, *J. Phys. A* **4**, 377, 564 (1971).
6. L. Allen and G. I. Peters, *J. Phys. A* **5**, 546, 695 (1972).
7. R. London, "The Length Dependence of Laser Power," unpublished (Lawrence Livermore National Laboratory, Livermore, California, Feb. 1987); also, C. J. Cerjan, private communication, February, 1987.
8. G. J. Linford, E. R. Peressini, W. R. Sooy and M. L. Spaeth, *Appl. Optics* **13**, 379 (1974).
9. K. Patek, "Lasers" (Iliffe Books, Ltd., London, 1964).
10. H. R. Griem, "Plasma Spectroscopy" (McGraw Hill, New York, 1964).
11. H. R. Griem, "Spectral Line Broadening by Plasmas" (Academic Press, New York, 1974).
12. R. C. Elton, *Appl. Optics* **14**, 2243 (1975).
13. R. W. Waynant and R. C. Elton, *Proc. IEEE* **64**, 1059 (1976).
14. O. S. Heavens, "Optical Masers" (Methuen Publ., London, 1965).
15. G. R. Fowles, "Introduction to Modern Optics" (Holt, Rinehart and Winston Publ., New York, 1975).
16. R. C. Elton, in "Methods of Experimental Physics, Plasma Physics," Vol. 1, Chapt. 4, H. R. Griem and R. H. Lovberg, eds. (Academic Press, New York, 1970).
17. H. R. Griem, *Phys. Rev. A* **33**, 3580 (1986).
18. L. W. Casperson, *J. Appl. Phys.* **48**, 256 (1977); also J. C. Camparo, *Phys. Rev. A* **39**, 69 (1989) on power broadening.
19. D. Marcuse, *Proc. IEEE* **51**, 849 (1963).

20. H. K. Holt, *Phys. Rev. A* **16**, 1136 (1977).
21. V. G. Arkhipkin and Yu. I. Heller, *Phys. Letters* **98A**, 12 (1983).
22. S. E. Harris, *Phys. Rev. Letters* **62**, 1033 (1989).
23. O. Svelto, "Principles of Lasers" (Plenum Press, New York, 1976).
24. D. A. Eastham, "Atomic Physics of Lasers" (Taylor and Francis, London and Philadelphia, 1986).
25. N. M. Ceglio, D. G. Stearns, D. P. Gaines, A. M. Hawryluk and J. E. Trebes, *Optics Letters*, **13**, 108 (1988).
26. A Illyukhin, G. Peregudov, E. Ragozin, I. Sobelman and V. Chirkov, *JETP Letters* **25**, 535 (1977).
27. N. M. Ceglio, D. P. Gaines, J. Trebes, A. M. Hawryluk, D. G. Stearns and G. L. Howe, *SPIE PROC.* **688**, 44 (1986).
28. W. L. Bond, M. A. Duguay and P. M. Rentzepis, *Appl. Phys. Lett.* **10**, 216 (1967).
29. R. M. J. Cotterill, *Appl. Phys. Letters* **12**, 403 (1968).
30. R. D. Delattes, *Appl. Phys. Letters* **12**, 133 (1968).
31. A. V. Kolpakov, R. N. Kuzmin and V. M. Ryaboy, *J. Appl. Phys.* **41**, 3549 (1970).
32. R. A. Fisher, *Appl. Phys. Letters* **24**, 598 (1974).
33. A. Yariv, *Appl. Phys. Letters* **25**, 105 (1974).
34. C. Elachi, G. Evans and F. Grunthaner, *Appl. Opt.* **14**, 14 (1975).
35. G. A. Lyakhov, *Sov. J. Quant. Electron.* **6**, 456 (1976).
36. R. W. P. McWhirter, in "Plasma Diagnostic Techniques," R. H. Huddleston and S. I. Leonard, p. 234 (Academic Press, New York, 1965).
37. R. J. Weymann and R. E. Williams, *Astrophys. J.* **157**, 1201 (1969).
38. D. G. Hummer and G. B. Rybicki, *Astrophys. J.* **254**, 767 (1982).
39. T. F. Stratton, in "Plasma Diagnostic Techniques," R. H. Huddleston and S. I. Leonard, pp. 391–393 (Academic Press, New York, 1965).
40. W. L. Wiese, M. W. Smith and B. M. Glennon, "Atomic Transition Probabilities, Vol. 1: Hydrogen through Neon," NSRDS-NBS-4 (U.S. Government Printing Office, Washington, DC, 1966).
41. V. V. Sobelev, *Sov. Astron. AJ* **1**, 665 (1957).
42. G. J. Pert, *J. Phys. B* **9**, 3301 (1976).
43. B. L. Whitten, R. A. London and R. S. Walling, *J. Opt. Soc. Am. B* **5**, 2537 (1988).
44. R. A. London, "Line Escape Probabilities for Exploding Foils," in Laser Program Annual Report 86, M. L. Rufer and P. W. Murphy, eds., Univ. of California Report No. UCRL-50021-86 (Lawrence Livermore National Lab., Livermore, California, 1988).
45. R. C. Elton, *Phys. Rev. A* **38**, 5426 (1988).
46. R. C. Elton, "Reduction of Radiative Trapping Effects in X-Ray Lasers using Autoionizing Transitions," Memorandum Report No. 5906 (U.S. Naval Research Laboratory, Washington, DC, 1986).
47. R. C. Elton, "Autoionization for Lower Level Detrapping in X-Ray Lasers," Formal Report No. 9103 (U.S. Naval Research Laboratory, Washington, DC, 1988).
48. J. G. Lunney, *Optics Comm.* **53**, 235 (1985).
49. J. P. Apruzese and J. Davis, *Phys. Rev. A* **31**, 2976 (1985).
50. R. C. Elton, T. N. Lee and W. A. Molander, *J. Opt. Soc. Am. B* **4**, 539 (1987).
51. J. C.-Riordin, J. S. Pearlman, M. Gersten and J. E. Rauch, in AIP Conf. Proc. No. 75, p. 34, D. Attwood and B. Henke, eds. (American Physical Society, New York, 1981); also M. Gersten, et al., *Phy. Rev. A* **33**, 477 (1986).
52. V. A. Chirkov, *Sov. J. Quantum Electron.* **14**, 1497 (1984).
53. M. D. Rosen, et al., *Phys. Rev. Letters* **54**, 106 (1985).
54. D. L. Matthews, et al., *J. de Physique*, Colloque C6, **47**, C6-1 (1986).
55. D. L. Matthews, et al., *Phys. Rev. Letters* **54**, 110 (1985).
56. J. P. Dahlburg, J. H. Gardner, M. H. Emery and J. P. Boris, *Phys. Rev. A* **35**, 2737 (1987).

57. "Advanced Technology Developments: 2.A Studies of New Geometries for X-Ray Lasers," in LLE Review **30**, 68 (University of Rochester Laboratory for Laser Engineering, Rochester, New York, 1987).
58. E. E. Fill, *Optics Comm.* **67**, 441 (1988).
59. J. D. Shipman, Jr., *Appl. Phys. Letters* **10**, 3 (1967).
60. R. W. Waynant, J. D. Shipman, Jr., R. C. Elton and A. W. Ali, *Appl. Phys. Letters* **17**, 383 (1970); and *Proc. IEEE* **59**, 679 (1971).
61. R. W. Waynant, *Phys. Rev. Letters* **28**, 533 (1972).
62. S. Suckewer, C. H. Skinner, H. Milchberg, C. Keane and D. Voorhees, *Phys. Rev. Letters* **55, 1753** (1985).
63. M. H. Key, *J. de Physique*, Colloque C1, **49**, C1-135 (1988).
64. R. C. Elton, *Comments At. Mol. Phys.* **13**, 59 (1983). Because of lack of printing clarity, there is some difficulty in distinguishing subscripts "ell" from "one" in this paper. The author will provide a corrected copy upon request.
65. R. W. P. McWhirter and A. G. Hearn, *Proc. Phys. Soc.* **82**, 641 (1963).
66. F. E. Irons and N. J. Peacock, *J. Phys. B* **7**, 1109 (1974).
67. R. C. Elton, T. N. Lee, R. H. Dixon, J. D. Hedden and J. F. Seely, in "Laser Interactions and Related Plasma Phenomena," H. Schwarz, H. Hora, M. Lubin and B. Yaakobi, eds., Vol. 5, p. 135 (Plenum Press, New York, 1981).
68. P. L. Hagelstein, "Physics of Short Wavelength Laser Design," Report No. UCRL-53100 (Lawrence Livermore National Laboratory, Livermore, California, 1981); and G. B. Zimmerman and W. L. Kruer, *Comments on Plasma Physics and Controlled Fusion* **2**, 51 (1975) (describes LASNEX); also Kim, et al., *J. Opt. Soc. Am. B* **6**, 115 (1989).
69. V. L. Jacobs, J. Davis, J. E. Rogerson and M. Blaha, *J. Quant. Spectros. and Radiative Transfer* **19**, 591 (1978).
70. K. W. Hill, M. Bitter, D. Eames, S. von Goeler, N. R. Sauthoff and E. Silver, in AIP Conference Proceedings No. 75, "Low Energy X-Ray Diagnostics," D. T. Atwood and B. L. Henke, eds., p. 10 (American Institute of Physics, New York, 1981).
71. S. Suckewer and H. Fishman, *J. Appl. Phys.* **51**, 1922 (1980).
72. S. Suckewer, C. H. Skinner, H. Milchberg, C. Keane and D. Voorhees, *Phys. Rev. Letters* **55**, 1753 (1985).
73. C. Keane, et al., *Bull. Am. Phys. Soc.* **33**, 2041 (1988).
74. M. C. Richardson, R. Epstein, O. Barnouin, P. A. Jaanimagi, R. Keck, H. G. Kim, R. S. Marjoribanks, S. Noyes, J. M. Sources and B. Yaakobi, *Phys. Rev. A* **33**, 1246 (1986).
75. T. N. Lee, E. A. McLean & R. C. Elton, *Phys. Rev. Letters* **59**, 1185 (1987).
76. H. Shiraga, et al., *Bull. Am. Phys. Soc.* **33**, 1947 (1988).
77. G. Jamelot, A. Klisnick, A. Carillon, H. Guennou, A. Sureau and P. Jaegle, *J. Phys. B* **18**, 4647 (1985).
78. M. H. Key, *Plasma Physics and Controlled Fusion* **26**, 1383 (1984).
79. P. Jaegle, *La Recherche* **18**, 16 (1987); and G. Jamelot, et al., in Proc. International Symposium on Short Wavelength Lasers and their Applications, C. Yamanaka, ed., p. 75 (Springer-Verlag, Tokyo, 1988).
80. I. N. Ross, J. Boon, R. Corbett, A. Damerell, P. Gottfeldt, C. Hooker, M. H. Key, G. Kiehn, C. Lewis and O. Willi, *Appl. Optics* **26**, 1584 (1987).
81. N. R. Pereira and J. Davis, "X-Rays from Z-Pinches," Memorandum Report No. 6097, Dec. 15, 1987 (Naval Research Laboratory, Washington, DC 20375).
82. S. J. Stephanakis, J. P. Apruzese, P. G. Burkhalter, J. Davis, R. A. Meger, S. W. McDonald, G. Mehlman, P. F. Ottinger, and F. C. Young, *Appl. Phys. Letters* **48**, 829 (1986).
83. R. Dukart, S. L. Wong, D. Dietrich, R. Fortner and R. Stewart, in Proc. Fifth Intl. Conf. High Power Particle Beams (U. Calif., San Francisco California 1983).
84. R. E. Stewart, D. D. Dietrich, P. O. Egan, R. J. Fortner and R. J. Dukart, *J. Appl. Phys.* **61**, 126 (1987).

85. R. B. Spielman, D. L. Hanson, M. A. Palmer, M. K. Matzen, T. W. Hussey and J. M. Peek, *J. Appl. Phys.* **57,** 830 (1985).

86. S. M. Zakharov, A. A. Kolomenskii, S. A. Pikuz and A. I. Samokhin, *Sov. Tech. Phys. Letters* **6,** 486 (1980).

87. R. H. Dixon and R. C. Elton, *Phys. Rev. Letters* **38,** 1072 (1977).

88. R. C. Elton, *Opt. Engr.* **21,** 307 (1982).

89. R. C. Elton, T. N. Lee and E. A. McLean, *J. de Physique*, Colloque C9, **48,** 359 (1987).

90. D. Jacoby, G. J. Pert, L. D. Shorrock and G. J. Tallents, *J. Phys. B* **115,** 3557 (1982).

91. P. Jaegle, G. Jamelot, A. Carillon, A. Sureau and P. Dhez, *Phys. Rev. Letters* **33,** 1070 (1974).

92. D. Matthews, et a., *J. Opt. Soc. Am. B* **4,** 575 (1987).

93. W. T. Silfvast and O. R. Wood, II, "Threshold for Laser Gain in the Presence of a Mirror Cavity" (unpublished).

94. "X-Ray and Gamma-Ray Imaging Techniques," D. Ramsden, W. B. Gilboy. R. P. Parker and A. J. Dean, eds., *Nucl. Instr. and Methods in Phys. Research* **221,** 1–187 (1984).

95. G. K. Skinner, ibid, p. 33; G. K. Skinner and T. J. Ponman, *J. British Interplanetary Soc.* **40,** 169 (1987); and G. K. Skinner, *Scientific American* **259,** 84 (August, 1988).

96. G. Schmahl, D. Rudolph and B. Niemann, in "Low Energy X-ray Diagnostics—1981," AIP Conference Proceedings No. 75, D. T. Attwood and B. L. Henke, eds., p. 225 (American Institute of Physics, New York, 1981).

97. N. M. Ceglio, ibid, p. 210.

98. D. C. Shaver, D. C. Flanders, N. M. Ceglio, and H. I. Smith, *J. Vac. Sci. Technol.* **16,** 1626 (1979); and H. J. Smith, E. H. Anderson, A. M. Hawryluk and M. L. Schattenburg, in "X-Ray Microscopy," G. Schmahl and D. Rudolph, eds., p. 52 (Springer-Verlag, New York, 1984).

99. H. I. Smith, in "Low Energy X-ray Diagnostics—1981," AIP Conference Proceedings No. 75, D. T. Attwood and B. L. Henke, eds., p. 223 (American Institute of Physics, New York, 1981).

100. A. N. Broers, W. W. Molzen, J. J. Cuomo and N. D. Wittels, *Appl. Phys. Letters* **29,** 596 (1976).

101. D. F. Parsons, ed., *Annals New York Acad. Sci.* **342,** (1980).

102. P. Kirkpatrick and A. V. Baez, *J. Opt. Soc. Am.* **38,** 766 (1948).

103. R. H. Price, in "Low Energy X-ray Diagnostics—1981," AIP Conference Proceedings No. 75, D. T. Attwood and B. L. Henke, eds., p. 189 (American Institute of Physics, New York, 1981).

104. W. Priedhorsky, ibid, p. 332.

105. J. A. R. Samson, "Techniques of Vacuum Ultraviolet Spectroscopy" (Pied Publications, Lincoln, Nebraska, 1967).

106. E. Källne, AIP Conference Proceedings No. 75, D. T. Atwood and B. L. Henke, eds., p. 97 (American Institute of Physics, New York, 1981).

107. W. R. Hunter, *SPIE Proceedings* **140,** 122 (1978).

108. A. M. Hawryluk, N. M. Ceglio, R. H. Price, J. Melngailis and H. I. Smith in AIP Conference Proceedings No. 75, "Low Energy X-ray Diagnostics," D. T. Atwood and R. L. Henke, eds., p. 286 (American Institute of Physics, New York, 1981).

109. M. C. Hettrick, in *SPIE Proceedings* **560,** 96 (1985).

110. T. Harada and T. Kita, *Appl. Optics* **23,** 3987 (1980).

111. M. C. Hettrick and S. Bowyer, *Appl. Optics* **22,** 3921 (1983).

112. N. Nakano, H. Kuroda, T. Kita and T. Harada, *Appl. Optics* **23,** 2386 (1984).

113. T. Kita, T. Harada, N. Nakano and H. Kuroda, *Appl. Optics* **22,** 512 (1983).

114. M. C. Hettrick, J. H. Underwood, P. J. Batson and M. J. Eckart, *Appl. Optics* **27,** 200 (1988).

115. L. S. Birks, "X-Ray Spectrochemical Analysis," second edition (Interscience Publ., New York, 1969).

116. I. W. Ruderman, K. J. Ness and J. C. Lindsay, Appl. Phys. L etters **7,** 17 (1965).

117. T. Arai, T. Shoji and R. W. Ryon, *Advances in X-Ray Analysis*, C. S. Barrett, P. K. Predecki and D. E. Leyden, eds., **28**, p. 137 (Plenum Press, New York, 1985).
118. J. A. Nicolosi, J. P. Groven, D. Merlo and R. Jenkins, *Opt. Engr.* **25**, 964 (1986).
119. N. J. Peacock, in AIP Conference Proceedings No. 75, "Low Energy X-ray Diagnostics," D. T. Atwood and R. L. Henke, eds., p. 101 (American Institute of Physics, New York, 1981).
120. B. L. Henke, F. G. Fujiwara, M. A. Tester, C. H. Dittmore and M. A. Palmer, *J. Opt. Soc. Am. B* **1**, 828 (1984).
121. W. Gudat, in "Uses of Synchrotron Radiation in Biology," H. B. Stuhrmann, ed., p. 32 (Academic Press, New York, 1982).
122. R. W. Waynant and R. C. Elton, in "Organic Scintillators and Liquid Scintillation Counting," p. 467 (Academic Press, New York, 1971).
123. J. L. Wiza, *Nuclear Instruments and Methods* **162**, 587 (1979).
124. M. J. Eckart and N. M. Ceglio, in Energy and Technology Review, November 1985, p. 25 (Lawrence Livermore National Laboratory, Livermore, California).
125. A. P. Lukirskii, E. P. Savinov, I. A. Brytov and Yu. F. Shepelev, *USSR Acad. Sci, Bull. Phys.* **28**, 774 (1964).
126. J. P. Henry, E. M. Kellogg, U. G. Briel, S. S. Murray, L. P. Van Speybroeck and P. J. Bjorkholm, *Proc. SPIE* **106**, 196 (1977).
127. T. Kondoh, et al., *Rev. Sci. Instrum.* **59**, 252 (1988); also N. Yamaguchi, et al., *Rev. Sci. Instrum.* **60**, 368 (1989).
128. D. J. Nagel, in "Low Energy X-Ray Diagnostics," AIP Conference Proceedings No. 75, D. T. Atwood and B. L. Henke, eds., p. 74 (American Institute of Physics, New York, 1981).
129. R. P. Madden and S. C. Ebner, ibid, p. 1.
130. R. C. Elton and L. J. Palumbo, *Phys. Rev. A* **9**, 1873 (1974).
131. J. Z. Klose and W. L. Wiese, "Extensions and an Expanded Data Set for the Branching Ratio Technique for Vacuum-UV Radiance Calibration," *J. Quant. Spectrosc. and Radiat. Transfer* (in press, 1989).
132. R. C. Elton, E. Hintz and M. Swartz, Proc. 7th International Conf. on Phenomena in Ionized Gases, Vol. 3, p. 190 (Gradevinska Knjiga Publ. House, Belgrade, Yugoslavia, 1966); also R. C. Elton and N. V. Roth, *Appl. Optics* **6**, 2071 (1967).

Chapter 3

Pumping by Exciting Plasma Ions

In this and the following three chapters, various pumping mechanisms will be described. Simple analyses are included, based largely upon relations developed in Chapter 2. The purpose of these analyses is four-fold:

(a) to illustrate the particular mechanism,
(b) to provide convenient scaling laws for extrapolation to shorter wavelengths,
(c) as examples of analytical techniques useful in the development of new schemes, and
(d) as an initial screening and a precursor to more costly and time consuming numerical modeling.

Experience has shown that ultimately it is the experimental laser research that provides the discrete and stepwise milestones which then feedback to the modeling for broader understanding and further projections. Hence, each analysis section here is followed by one summarizing existing experimental evidence for the particular approach. The results of Chapters 3–5 are summarized at the beginning of Chapter 7.

Analyses such as demonstrated here often lead to optimistic values of gain. This is largely because only a few levels and rates can be included. Hence, if a new scheme evaluated in this fashion promises only threshold gain, it probably is not feasible for an experiment in which interactions with other nearby levels takes place. This will become more apparent as the analyses progress.

In the excitation mode of pumping, the upper-laser-state is populated from the ground state of the same ion. Excitation may be induced either by the inelastic impact of a free electron or by photon absorption. The former will be discussed in this and Section 2 following. The latter is covered in

Sections 3 and 4. For each, an analysis section provides some useful analytical scaling relations. These are followed by a section on experimental results.

1. ANALYSIS OF ELECTRON-COLLISIONAL EXCITATION PUMPING

It seems prudent to advance into the x-ray spectral region by extrapolating from experience in the visible and near-ultraviolet regions. Following established procedures from solar and plasma spectroscopy, we can proceed iso-electronically (i.e., towards heavier elements with a constant number of bound electrons) to shorter wavelength transitions associated with an increased core charge. Electron-collisional excitation pumping is a prime example of such a systematic approach.

Electron-collisional excitation is perhaps the easiest x-ray laser pumping concept to understand from experience with ion lasers at long wavelengths. Thus it is not surprising that early analyses of this approach for the xuv spectral region were based on numerous six-electron C-like ion lasers operating sucessfully in the ultraviolet spectral region[1,2].

This method of pumping is described by the equation

$$X_o^{i+} + e \rightarrow X_u^{i+}. \tag{1}$$

Here X^{i+} represents an i-times ionized atom of element X in which pumping takes place from state o to the excited upper state u in the same species. This is also illustrated in the energy-level diagram of Fig. 1. In this case the initial

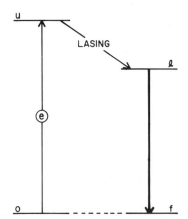

Fig. 1.   Energy level diagram for electron-collisional excitation pumping and lasing.

(o) and final (f) states are the same. Hence, the initial state population is immediately replenished by decay from the lower laser state.

1.1. General Relations

We begin the analysis with a simple but quite general and useful expression[3-5] for the electron-collisional excitation pumping rate for promoting a bound electron over an energy gap ΔE_{ou} from level o to a higher level u. This is a modified Coulomb-Born formula for electric dipole transitions from threshold to higher energies. Bethe-type corrections are included through a semi-empirical Kramers-Gaunt factor. The expression for the pumping rate P_{ou} (designated here as P_{ce} for collisional excitation) is

$$P_{ce} = N_e C_{ou}$$

$$= 64\pi^2 \left(\frac{\pi}{3}\right)^{1/2} \frac{Ry}{h} a_o^3 N_e \left(\frac{Ry}{kT}\right)^{1/2} \left(\frac{Ry}{\Delta E_{uo}}\right)$$

$$\times f_{ou} \langle g_{ou} \rangle \exp\left(-\frac{\Delta E_{uo}}{kT_e}\right) \quad \text{sec}^{-1}, \tag{2a}$$

or, for ΔE_{uo}, kT_e in eV,

$$P_{ce} = 1.6 \times 10^{-5} \frac{N_e f_{ou} \langle g_{ou} \rangle}{\Delta E_{uo}(kT_e)^{1/2}} \exp\left(-\frac{\Delta E_{uo}}{kT_e}\right) \quad \text{sec}^{-1}. \tag{2b}$$

Here, in general, $C_{ou} = \langle \sigma_{ou} v \rangle_{\text{maxw}}$ is the excitation rate coefficient averaged over a Maxwellian distribution function at an electron temperature T_e, for a cross-section σ_{ou} and velocity v. Also, f_{ou} is the absorption oscillator strength of the transition[6] and $\langle g_{ou} \rangle$ is the effective Gaunt factor averaged over a Maxwellian velocity distribution. This Gaunt factor is plotted[5] in Fig. 2 and is seen to be ≈ 0.2 for many ions, at least when $n_u - n_o \geq 1$. The formulae in Eqs. (2) are expected to be accurate to within a factor of two, based on numerous experiments in plasmas[7].

From Eqs. (2) it is noticed that the dependence of the pumping product $N_o P_{ce}$ [Eq. (19), Chapter 2] on electron density is to the second power (since N_o is proportional to N_e). This definitely favors high density. Also to be noted is that the lower laser level with a smaller energy gap ΔE_{ol} is more rapidly excited in an o-to-l transition than the upper (o-to-u). Hence, the lower level must depopulate more rapidly to achieve a quasi steady-state (cw) population inversion. This is guaranteed as long as the upper state remains sufficiently metastable against collisional as well as radiative decay.

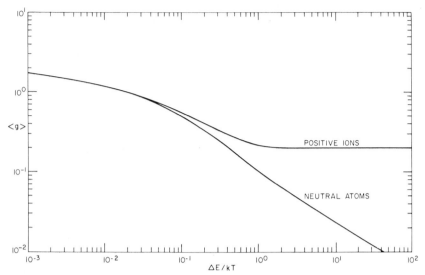

Fig. 2. The effective excitation Gaunt factor $\langle g \rangle$ averaged over a Maxwellian velocity distribution as a function of the ratio of the energy gap to the electron temperature. (From Ref. 5.)

We can use the general $N_e C$ relation in Eqs. (2) also for collisional mixing between the lower (l) and upper (u) laser levels, in establishing a maximum and an optimum value for the electron density (see Sections 2.4.2 of Chapter 2 and 1.2.1b below). For this we need the deexcitation rate $N_e C_{u\ell}$ from the upper (u) to the lower (ℓ) laser state. This is related to the excitation rate $N_e C_{\ell u}$ for these two levels through detailed balancing[5] by

$$ N_e C_{u\ell} = N_e C_{\ell u} \left(\frac{g_\ell}{g_u} \right) \exp\left(\frac{\Delta E_{u\ell}}{kT_e} \right). \tag{3} $$

Upon substituting from Eq. (2) for this particular u–ℓ transition, we have

$$ N_e C_{u\ell} = 1.6 \times 10^{-5} \frac{f_{\ell u} \langle g_{\ell u} \rangle N_e}{\Delta E_{u\ell}(kT_e)^{1/2}} \frac{g_\ell}{g_u}, \tag{4} $$

where g_ℓ and g_u are the statistical weights of the respective levels (not to be confused with the Gaunt factor). Here again $\Delta E_{u\ell}$ and kT_e are expressed in eV. Notice that the exponential factor cancels with this substitution.

Electron-collisional excitation pumping is not particularly suitable for $n = 3$ to $n = 2$ lasing in hydrogenic (or helium-like) ions[2]. Hence, the particularly simple hydrogenic scaling relations used for most of the other examples which follow cannot be used to describe a promising electron-collisional excitation pumping system. This is primarily because the hydrogenic $n = 3$ level is not metastable and is pumped at a slower rate than is the $n = 2$ level at the temperatures required. In fact, even for the hydrogenic low-temperature recombination pumping described in the Chapter 4, there is an upper cutoff[8] for temperature at a value of about $kT_e = 0.1 \times Z^2$ Ry for hydrogenic ions (see also Section 3.1 and Figs. 17 and 18 of Chapter 2). This cutoff is due to excessive excitation pumping of the lower ($n = 2$) level. This maximum temperature would be excessively low for significant electron-collisional excitation pumping.

1.2. Monopole Excitation of a Metastable Level

On the other hand, a very successful laser system based on electron-collisional pumping does exist. In its simplest form it involves 2p → 3p valence-electron excitation, where the 2p level is the ground state in a $1s^2 2s^2 2p^n$ configuration. The 3p level is quite metastable against direct spontaneous dipole decay to the ground state. However, the 2p → 3p monopole excitation rate is comparable to that for the 2p → 3s or 2p → 3d dipole rates, which in turn can be estimated from Eqs. (2). Lasing then takes place on a $\Delta n = 0$ transition from the 3p to the 3s level. The lower 3s state then rapidly depopulates back to 2p. Such transitions are closely spaced, such that heavy ions of much higher charge state are required to achieve the same lasing wavelengths as, for example, for hydrogenic $\Delta n = 1$ transitions.

1.2.1. Neon-like Ion Lasing

While any $2p^n$ ($n = 1$ to 6) isoelectronic sequence can be used in principle[1], the $n = 6$ neon-like sequence[9] has proven to be the most successful. This is at least partially because of a greater stability of the species in a transient plasma, which is associated with the higher ionization potential of the closed shell configuration (see Chapter 2, Section 3.2.2a and Fig. 19). The primary neon-like energy levels and configurations involved in the x-ray laser are shown in Fig. 3.

The notation used in the present analysis is that of LS-coupling. This is likely to be the most familiar to the reader and the easiest to relate to the other analyses which follow. For $Z \gtrsim 20$ however, considerable mixing occurs between the levels. Then, the true ionic structure is more clearly described by either jj or jk coupling. Both LS and jj designations for the six most prominent laser transitions are included later in Fig. 5.

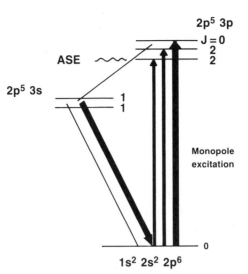

$2p^5\ 3d$

$2p^5\ 3p$

ASE

$2p^5\ 3s$

Monopole
excitation

$1s^2\ 2s^2\ 2p^6$

Fig. 3. Energy level diagram specifically for a Ne-like 3p-3s laser excited from the $2p^6$ ground state by electron collisions.

For neon-like ions, many sublevels exist in the $1s^2 2p^5 3p$ upper and $1s^2 2p^5 3s$ lower laser configurations. This is an added complexity compared to the simpler scaling possible with hydrogenic systems used for most examples in this book. Because it is generally agreed that the dominant $2p \rightarrow 3p$ excitation occurs on the $2p^6\ ^1S_0$-$2p^5 3p\ ^1S_0$ ($J = 0$) transition, followed by lasing on a transition to the $2p^5 3s\ ^1P_1$ lower-laser level, this will be used in the present analysis (see Fig. 3). It is now known from experiments that two other transitions originating on $J = 2$ levels namely $3p\ ^3P_2$-$3s\ ^1P_1$ and $3p\ ^1D_2$-$3s\ ^3P_1$ show higher gain. However, the exact pumping channels and plasma conditions are not yet fully understood. (The evidence and some possible explanations for this are discussed in Section 2.)

In the Ne-like ion analysis which follows, we establish a general procedure to be continued in later sections for other pumping schemes. We begin by deriving an expression for the collisional excitation rate C_{ou} as a function of the net charge $\zeta = Z - 9$ influencing the active electron, where Z is the nuclear charge. Next we derive a relation for the optimum electron density $(N_e)_{opt}$ for the lasing system, first in terms of $Z - 9$ and then of the laser wavelength $\lambda_{u\ell}$. From these we obtain a wavelength-scaling expression for the pumping rate $P_{ce} = N_e C_{ou}$. Finally, we express the upper state density N_u in terms of the initial-state density N_o, taking account of collisional as well as radiative depopulation of level u, and obtain an estimate of the scaling of the gain coefficient G_{ce} with laser wavelength.

Throughout the present analysis we will be evaluating the relations developed for the particular neon-like Se^{24+} ($Z = 34$) ion. This element is somewhat of an optimum choice for high gain[10] as well as a very successful species used in laser experiments near 200 Å (see Section 2). Hence, extensive theoretical as well as experimental data are available for normalization in this analysis. Also, this is a convenient wavelength region for later comparison with hydrogenic C^{5+} Balmer-α examples at 182 Å and others nearby, discussed elsewhere in this book. (Chapters 3–5)

1.2.1a. THE COLLISIONAL-EXCITATION RATE COEFFICIENT. Calculations[11,12] have shown that the 2p-3p monopole excitation rate is approximately equal to one-half that of the 2p-3d dipole rate, which we can estimate from Eqs. (2) above. For such an estimate, we need values and Z-scaling relations for several parameters. To begin, an electron temperature of $kT_e = \chi/3$ is assumed, where χ is the ionization potential. Atomic physics data[11] further indicate that we can scale both this ionization poetntial as well as the 2p → 3p excitation energy ΔE_{uo} in a similar 1.5 power law of the net charge $Z - 9$, i.e., empirically

$$\chi = 3kT_e \approx 17(Z - 9)^{1.5} \quad \text{eV}, \tag{5}$$

and

$$\Delta E_{uo} \approx 11(Z - 9)^{1.5} \quad \text{eV}. \tag{6}$$

Equations (5) and (6) then give $\Delta E_{ou}/kT_e = 1.9$, from which a constant Gaunt factor $\langle g_{ou} \rangle = 0.2$ is obtained, using Fig. 2. The same data indicate that a mean value of

$$f_{ou} \approx 1.6 \tag{7}$$

is appropriate for the 2p-3d oscillator strength. (These scaling relations are valid at least for moderate-Z ($= 20$–36) elements.)

Substituting these parameters into Eq. (2b) and dividing by two yields for the 2p → 3p $^1S_0$ excitation rate coefficient

$$C_{ou} = \frac{1.5 \times 10^{-8}}{(Z - 9)^{2.25}} \quad \text{cm}^3 \text{ sec}^{-1}. \tag{8}$$

For selenium, $Z - 9 = 25$, and this gives $C_{u\ell} = 1.0 \times 10^{-11} \text{ cm}^3 \text{ sec}^{-1}$. This value compares favorable with a value obtained from relativistic distorted wave calculations[12] of $1.02 \times 10^{-11} \text{ cm}^3 \text{ sec}^{-1}$ and a value[13] of

1.2×10^{-11} cm^3 sec^{-1} used in describing the initial selenium experiments. This adds confidence in the present simple analysis. Next we will derive a Z-scaling value for the electron density.

1.2.1b. OPTIMUM ELECTRON DENSITY. For the electron density, we adopt an optimum value $(N_e)_{opt}$ equal to one-half[9] of $(N_e)_{max}$. A general expression for this maximum value was derived in Eqs. (25) and (26) of Chapter 2. That derivation was based on the maximum collisional deexcitation rate permissible for the laser transition, according to Eqs. (2)–(4) above. For 3p-3s $\Delta n = 0$ laser transitions in neon-like ions, the validity of this Bethe-Born formulation is somewhat questionable, at least using $\langle g \rangle$ values shown in Fig. 2. Hence, before using it here (mainly for consistency), we can compare the excitation rate coefficient $C_{\ell u}$ obtained with Eqs. (2) for $Z = 34$ (selenium) with accurate relativisitic distorted-wave calculations[12]. From this, we arrive at an effective Gaunt factor approximately equal to unity. In an alternate method of estimating this rate[1], the theoretical line width (in frequency units) is used. The expression involved is somewhat unwieldy for convenient scaling here. However, comparison with the result of Eqs. (2) above indicates an effective Gaunt factor of two for selenium. Based on these informations, we adopt a constant effective Gaunt factor of $\langle g_{u\ell} \rangle_{eff} \approx 1.5$. This increase in Gaunt factor for $\Delta n = 0$ transitions is in agreement with a host of other excitation data[7].

For the ratio of transition probabilities in Eq. (25) and (26) of Chapter 2, we again rely upon theoretical calculations[11,13] for guidance and adopt

$$\frac{A_{\ell o}}{A_{u\ell}} \approx 100, \tag{9}$$

at least for medium-Z elements. From the same theoretical results[11], we take

$$\Delta E_{u\ell} = 2.7(Z - 9) \quad \text{eV} \tag{10}$$

for the present laser transition. This latter relation also gives the generally useful conversion

$$\lambda_{u\ell} = \frac{4600}{Z - 9}, \tag{11}$$

for the x-ray laser wavelength in Angstrom units again.

The maximum value for the electron density in a neon-like x-ray laser plasma then numerically is equal to

$$(N_e)_{max} = 8.3 \times 10^{15} (Z - 9)^{3.75} \quad \text{cm}^{-3}, \tag{12}$$

The scaling power of Z shown here is less than for hydrogenic ions($\propto Z^7$) given in Eq. (28) of Chapter 2. This is due in part to the linear scaling of wavelength here [Eq. ⌐◡) above], compared with quadratic for hydrogenic ions. While the maximum densities are comparable for Ne -like selenium and hydrogenic carbon, both lasing near 200 Å, this difference in scaling makes a considerable difference in the optimum density that can be used for shorter wavelengths.

The optimum value of electron density is then given by one-half of the relation in Eq. (12), i.e.,

$$(N_e)_{opt} = 4 \times 10^{15} (Z - 9)^{3.75} \quad cm^{-3}. \tag{13}$$

Converting to laser wavelength using Eq. (11), this becomes

$$(N_e)_{opt} = \frac{2 \times 10^{29}}{\lambda_{u\ell}^{3.75}} \quad cm^{-3}, \tag{14}$$

for $\lambda_{u\ell}$ in Angstrom units. Again using as an example Ne-like selenium at a wavelength of 207 Å corresponding to the strong lasing lines observed in experiments, we derive from this simple analysis $(N_e)_{opt} \approx 4 \times 10^{20}$ cm^{-3}. This is essentially the same value reported[13] for the experiments.

1.2.1c. THE PUMPING RATE.   Combining Eq. (14) with Eq. (8) (converted to laser wavelength), the pumping rate $P_{ce} = N_e C_{ou}$ becomes

$$P_{ce} = \frac{2 \times 10^{13}}{\lambda_{u\ell}^{1.5}} \quad sec^{-1}. \tag{15}$$

For comparison, we can take $\lambda_{u\ell} = 207$ Å again, and derive

$$P_{ce}(Se) = 7 \times 10^9 \quad sec^{-1}. \tag{16}$$

This is in close agreement with a value of 5.8×10^9 sec^{-1} assumed for pumping the dominant transition in the experiment[13].

1.2.1d. THE UPPER-STATE DENSITY.   From Eq. (19) of Chapter 2, we can relate N_o to the upper-laser-state density N_u by

$$N_u = N_o \frac{P_{ce}}{A_{ul} + C_{ud}}. \tag{17}$$

Here we have included collisional deexcitation of the upper level to nearby 3d levels through C_{ud}, by substituting $D_u = A_{u\ell} + C_{ud}$.

The density of initial neon-like ground-state ions N_o is related to that of the electrons by $N_o = N_e/3(Z - 10)$, which is taken for convenience to be approximately

$$N_o = \frac{N_e}{3(Z-9)} = \frac{N_e \lambda_{u\ell}}{14000} \quad cm^{-3}, \tag{18}$$

at moderate Z and again for $\lambda_{u\ell}$ in Angstrom units. The 1/3 here represents that portion of the total ion density in the neon-like ground state[10].

In Eq. (17), C_{ud} may be estimated from the collisional depopulation rate out of the upper level into the higher 3d $^1P_1$ level (shown in Fig. 3), which in turn decays rapidly to the $2p^6$ $^1S_0$ ground state. A value for this rate is obtained from Eqs. (2) and (14) above. Again, we use a $\Delta n = 0$ Gaunt factor of 1.5, $f_{ud} = 0.16$ and

$$\Delta E_{ud} = 2.5(Z-9) = \frac{11500}{\lambda_{u\ell}} \quad eV, \tag{19}$$

from atomic data[11]. The result is

$$N_e C_{ud} = \frac{5 \times 10^{16}}{\lambda_{u\ell}^2} \quad sec^{-1}. \tag{20}$$

Comparing this to the radiative rate $A_{u\ell}$ found from Eqs. (11) of Chapter 2, using[11] $f_{\ell u} = 0.040$ and $g_\ell/g_u = 3$, we have

$$N_e D_{ud} = 70 \, A_{u\ell}. \tag{21}$$

Hence, we can assume that $D_u = C_{ud}$ in Eq. (17) above, which then becomes simply

$$N_u = N_o \frac{P_{ce}}{N_e C_{ud}}. \tag{22}$$

From this and Eqs. (15) and (20) above, we obtain the degree of pumping

$$\frac{N_u}{N_o} = 3 \times 10^{-4} \, (\lambda_{u\ell})^{0.5}. \tag{23}$$

For selenium lasing at $\lambda_{u\ell} = 207$ Å, this becomes $N_u/N_o = 0.005$, which agrees with more detailed computations for the present temperature chosen[10].

Furthermore, using Eq. (9), we find from Eq. (21) that

$$N_e D_{ud} = 0.7 \times A_{\ell o}, \tag{24}$$

at an electron density of $N_e = (N_e)_{opt}$. Hence, the lower-laser-level radiative decay still dominates over the total decay rate of the upper level, and quasi-cw operation is assured.

Combining Eqs. (14) and (18), we derive

$$N_o = \frac{2 \times 10^{25}}{\lambda_{u\ell}^{2.75}} \quad cm^{-3}, \tag{25}$$

and, with Eq. (23)

$$N_u = \frac{6 \times 10^{21}}{\lambda_{u\ell}^{2.25}} \quad cm^{-3}. \tag{26}$$

For $\lambda_{u\ell} = 207$ Å, this becomes $N_u = 4 \times 10^{16} \, cm^{-3}$.

1.2.1e. THE GAIN COEFFICIENT.   Using Eqs. (8) and (12'') (Table 1) of Chapter 2 with substitution from Eq. (26), and other parameters given above, we arrive at a gain coefficient of

$$G_{ce} \approx \frac{6500}{\lambda_{u\ell}^{1.25}} \quad cm^{-1}, \tag{27}$$

for $\lambda_{u\ell}$ in Angstrom units and an inversion factor of $F = \frac{1}{3}$. The net increase in gain at shorter wavelengths is due to the rapid rise in density [Eq. (14)] off-setting a normal decrease in the gain coefficient. Hence, the scaling will be sensitive to that of the electron density for the particular model. For $\lambda_{u\ell} \approx 207$ Å, we now have $G_{ce} \approx 8 \, cm^{-1}$, which is close to that which is measured for selenium (Section 2).

1.2.2. Variations on This Concept

1.2.2a. NICKEL-LIKE IONS.   The above relations show that the gain does not extrapolate particularly rapidly with shorter wavelengths. The general scheme can however be carried further with Ni-like ions for which the laser transitions take place on $n = 4$, $\Delta n = 0$ transitions[14]. Shorter wavelengths are obtained using higher-Z atoms. It is predicted that such electron-collisionally pumped $\Delta n = 0$ lasing can then be extended to the very desirable 40 Å region. The gain coefficient is necessarily reduced somewhat however, because of the larger number of levels involved compared to the 3p-3s approach. Experimental results will be summarized in Section 2.

1.2.2b. ENHANCED PUMPING BY SUPRATHERMAL ELECTRONS.   The above ideas are based, as usual, on the assumption of a Maxwellian energy distribution for the electrons in the plasma. There have been suggestions that pumping could be enhanced by generating a non-Maxwellian velocity component[15,16]. This might consist of a 10% added high-energy component of "suprathermal" electrons of energy approximately matching the 2p → 3p excitation energy for neon-like ions. It is suggested that it may be possible to intentionally tailor such a distribution near the critical layer in laser-produced plasmas of sufficient irradiance. Enhancements in gain by factors as high as 10 times are predicted under ideal circumstances. Excitation pumping by electron beams apparently has not been seriously proposed, even though intense beams of current density $> 1 \text{ kA/cm}^2$ are possible[17].

1.2.2c. ALTERNATE PUMPING MODES FOR $\Delta n = 0$ TRANSITIONS.   From   the above analysis, it should not be concluded that electron-collisional excitation from the ground state is the only viable pumping method for $\Delta n = 0$ lasing. It was chosen here as an example which has been very successful experimentally (see Section 2 following). Recombination pumping, discussed in Chapter 4, is an attractive alternative, i.e., filling the excited states from above at a low temperature, rather than from below. Indeed it has been suggested that some of the experimental results at least can be explained by three-body collisional[18] as well as dielectronic recombination[19] processes, starting with a nine-electron F-like plasma for the Ne-like laser example.

2.   EXPERIMENTS INVOLVING ELECTRON-COLLISIONAL EXCITATION PUMPING

There are hundreds of electrical-discharge-driven lasers operating in the infrared through ultraviolet spectral regions[20]. Most of these are thought to be pumped by electron-collisional excitation. Some operate on neutral atoms and molecules. Others use ions in the first few stages as amplifying species.

2.1. Vacuum-UV Molecular Lasers

In progressing into the vacuum-uv spectral region of decreasing cavity efficiency and finally ASE operation, it is natural to begin with knowledge and experience gained in the near-uv.

2.1.1. Molecular Nitrogen and Hydrogen Lasers

The first example of this continuation began with the molecular nitrogen (N_2) laser. This medium was made to operate at an ultraviolet wavelength of 3371 Å in the ASE mode without a cavity, and with traveling-wave enhancement of the output power[21]. This was a major advance, paving the way

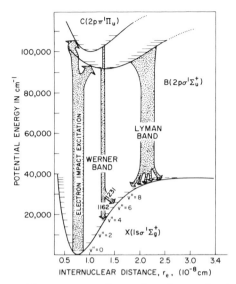

Fig. 4. Simplified potential energy versus internuclear distance for the hydrogen molecule. (From Ref. 25, © 1976, IEEE.)

for non-cavity vacuum-uv amplification. Subsequently, operation was extended[22−24] successfully into the vacuum-uv region in the 1100- and 1600-Å wavelength range using molecular hydrogen (H_2).

In order to better associate this device with electron-collisional excitation, we refer to the energy level diagram[25] for H_2, shown in Fig. 4. For this molecule the ground state is stable. At room temperature, only the ground vibrational level is filled. The $B(^1\Sigma_u^+)$ and $C(^1\pi_u)$ upper states shown are filled by electron-collisional excitation, according to the Franck-Condon factors and the energy spectrum of the free-electrons. Self-terminating lasing occurs on Lyman- and Werner-bands transitions to the $X(^1\Sigma_g^+)$ lower state in the 1600-Å and 1160-Å regions, for the B and C upper states, respectively. In all, 176 xuv-lasing lines in the 1098–1646 Å range have been reported for hydrogen and deuterium isotopes[20].

The actual 1-meter long ASE device used is described in Section 4.2.2a of Chapter 2. The pulse lengths measured were ≈1 ns in duration. The traveling-wave characteristic was evidenced by a greater than 10:1 enhancement in output intensity from the end towards which the excitation traveled, compared to the opposite end. The output power measured in this traveling-wave device was as high as 100 kW. Other versions of the H_2 laser operating at higher pressure and also repetitively on a smaller scale have been developed, as described in a review article[26].

2.1.2. Rare-Gas Excimer Lasers

Closely related to this are the vacuum-uv molecular rare gas excimer lasers[20,27], most notably Xe_2^*. This particular laser system operates[28] in a quasi-cw mode at 1716 Å. Here the molecule only exists in the excited state, and the ground state is unstable to rapid dissociation. Hence lasing is continuous (cw), as long as the necessary molecular density is maintained. The excited-state molecules are most efficiently produced by various excitation and ionization channels in high-pressure (100–500 psia) volumes of the rare gas[29]. Beams of electrons accelerated through 300 keV to 2 Mev are typically used as drivers. Other excimers, e.g., Ar_2^* and Kr_2^* also have been made to lase at wavelength of 1261 and 1457 Å, respectively[20,27].

2.2. Vacuum-UV Ion Lasing in a Capillary Discharge

Ultraviolet and vacuum-uv lasing on 10 lines spanning the 1756–2315 Å range has been demonstrated in a simple repetitive capillary discharge device[30]. Triply ionized rare-gas atoms such as neon, argon and krypton were typically used, at a pressure of 5–30 mTorr. Pumping presumably occurred from electron-collisional ionization followed by excitation. Peak powers as high as 1 kW were obtained. The dimensions were 7-mm diameter and 150-cm length. Cavities with reflectivities in the 96–98% range were used. Current densities reached 14 kA/cm^2 from energy stored in a 0.12 μF capacitor switched by a thyratron. The pulse length was 500 ns, with a <200 ns risetime. The repetition rate was multi-Hertz. Continuous operation for 10^9 discharges was possible without maintenance.

2.3. Xuv Plasma Ion Lasers

A fairly obvious way to systematically extend visible and uv ion lasers to shorter wavelengths is along isoelectronic sequences, i.e., with ions of progressively higher nuclear charge Z but with a constant number of bound electrons. The technology involving an electrical discharge in a highly reflecting cavity is already extendable to the 1000–2000 Å vuv spectral region. This is evidenced by the experiments reported above. However, a transition to densities increased by as much as 100 times is required as a result of a shift from cavity to non-cavity ASE operation at shorter wavelengths. This then becomes a plasma environment in which collisional mixing and radiative trapping play an increasingly important role as the wavelength becomes shorter.

Most long-wavelength ion lasers operate on $\Delta n = 0$ transitions, with $n = 3$ or 4 typical[20]. A particular wavelength is more easily achievable with a lower stage of ionization for $n = 3$ rather than $n = 4$. Hence the scaling analysis in Section 1 above was primarily based on $n = 3–3$ transitions. This is also true of experiments. Therefore, most of the discussion following revolves around successful 3p-3s lasing in electron collisionally pumped Ne-like ions.

2.3.1. Spontaneous Emission and Fluorescence

The measurement of fluorescence and population inversion by relative line intensities as a precursor to net gain (discussed in Section 5.1 of Chapter 2) is not as straightforward for the Ne-like sequence as for hydrogenic ions. This is because the upper state does not readily decay to the ground state, i.e., it is metastable. This of course is the primary feature of this scheme that permits excitation pumping with a broadband energy source such as a Maxwellian plasma electron environment. It is in principle possible to compare the 3p upper-laser-state population density found from 3p-3s xuv spontaneous emission with the 3s lower-laser-population density from a 3s-2p x-ray resonance transition. The former tends to be weak and the latter possibly optically thick, so that both require careful consideration. The relative calibration needed to span the large wavelength difference between the two spectral lines can, within similar limitations, be obtained from a branching ratio between 3d-3p and 3d-2p lines. (These techniques are also described in Sections 5.1 and 6.6.2b of Chapter 2.)

The intensities of 3p-3s and 3d-3p spontaneous transitions required for such fluorescent measurements (as well as for an analysis of the excited-state kinetics of the system) are generally very low. However, the wavelengths for these transitions have been observed for selected neon-like ions in beam-foil as well as in various plasma devices (references are compiled in Ref. 31). Even more recently, intensity measurements have been reported for Ar^{8+} ions from a well-diagnosed theta-pinch device[32]. However, the most success for developing this type of x-ray lasers has come from direct measurements of gain, rather than through evolution from fluorescence and population inversion measurements.

2.3.2. 3p-3s Gain Experiments in Ne-Like Ions

All of the successful $\Delta n = 0$ gain experiments to date have involved plasma media created by the transverse irradiation of planar targets with line-focused high-power laser beams. This approach is illustrated for multiple driving beams in Figs. 20 of Chapter 2 and on the cover.

The potential for lasing on Ne-like $n = 3$, $\Delta n = 0$ transitions was reportedly[33] first demonstrated in the xuv spectral region near 600 Å. The experiment was performed in a laser-produced calcium plasma generated from a solid slab target of width 30–125 μm wide and length 20–40 mm. The driver was a Nd:glass laser of wavelength 1 μm, energy 30 J, and duration 2.5–5 ns. It was focused to a spot of 400–800 μm width and 10–40 mm length for an irradiance of approximately 5×10^{10} W/cm^2. Included in the vacuum chamber was a resonant cavity consisting of two mirrors of $\sim 25\%$ and 13% reflectivities and an intermediate grating mounted in zero-order for grazing-incidence reflection. This grating served to divert (through dispersion) a portion of the laser beam. A high intensity in a narrow spectral zone near 600 Å was recorded photographically. This was associated with 3p-3s lasing in that region. These initial experiments apparently never were followed up or corroborated.

Significant 3p-3s gain at shorter (≈ 200 Å) wavelengths was demonstrated[34] convincingly in the mid-1980s at the Lawrence Livermore National Laboratory (LLNL), approximately 10 years after the conceptualization of this approach[1]. These experiments involved plasmas produced by a 0.53 μm, 1 kJ, 0.45 ns laser. This driving beam was line focused (100 μm by 11 mm) onto ultra-thin-film targets made of selenium (750-Å thick) deposited onto a formvar substrate[13]. A typical target irradiance was 5×10^{13} to 1×10^{14} W/cm^2. The initial experiments were also performed with yttrium Y ($Z = 39$). They were later extended to $Z = 42$ with 2000 J of laser energy, using molybdenum (Mo) coatings[35]. These experiments were continued in collaboration with the Centre d'Etude de Lemeil (CEL), France, and extended to a 1.05 μm driving-laser wavelength. They were also performed with strontium, with similar results[36]. A semi-popular overview of this series of experiments at LLNL can be found in Ref. 37.

At lower Z and lower irradiance (1.05 μm wavelength, 500 J, 2 ns), xuv lasing was demonstrated at the Naval Research Laboratory (NRL) with similar copper/formvar thin-film targets, as well as with much thicker (1.3 μm) unbacked copper foils[38]. Here the target irradiance was ~6 × 10^{12} W/cm^2. These experiments were also performed with solid targets of copper, germanium, zinc, and gallium-arsenide[38,39]. Solid-slab targets may prove advantageous over thin films for extended lasing times and multipass cavity operation.

2.3.3. Lasing Energy Levels and Wavelengths

In Ne-like ions, there are ten 3p levels and four 3s levels. Of the many possible transitions to the two 3s $J = 1$ levels (which rapidly depopulate radiatively to the ground state), lasing has been detected on the six which are diagrammed and labeled in Fig. 5. In this level diagram, both jj and LS coupling notations are shown. The former is the more accurate description for the Ne-like sequence. The latter is included mainly for convenience of discussion and for the reader's familiarity. The LS labels given in Fig. 5 represent the strongest term in each mixture for selenium. As just one example of possible ambiguities that can arise with the LS notation, the 3p $^3p_2$ and $^1D_2$ designations shown in Fig. 5 for selenium can reverse at lower Z and different term mixtures[32].

The measured 3p-3s wavelengths for transitions which have been shown to lase in Ne-like ions are collected in Table 1, according to the alphabetical labeling shown in Fig. 5. The wavelengths quoted to two decimal places are considered to be accurate[41] to ±0.04–0.06 Å. Except for the "A" transition which is difficult to extrapolate, these wavelengths agree[39-42] generally within ~0.1 Å with theoretical values[43-45].

Table 1

Measured Wavelengths [Å] of 3p-3s Lasing Lines in Ne-Like Systems.

Transition	A	B	C	D	E	F	Ref.
Species							
Cu^{19+}	221.11	279.31	284.67				38
Zn^{20+}	212.17	262.32	267.23				41
Ga^{21+}		246.70	251.11				39
Ge^{22+}	196.06	232.24	236.26	247.32	286.46		40
As^{23+}		218.84	222.56				41
Se^{24+}	182.43	206.38	209.78	220.28	262.94	169.29	36, 45
	182.44	206.35	209.73				41
Sr^{28+}	159.8	164.1	166.5				36
Y^{29+}		155.0	157.1	165	218		36, 42
Mo^{32+}	141.6	131.0	132.7	139.4		106.4	35, 36

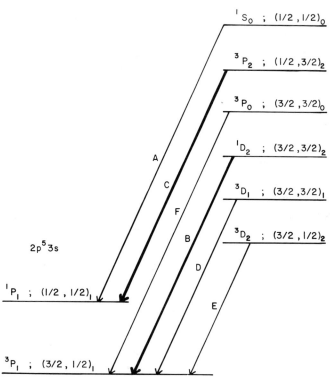

$2p^5 3p$

$^1S_0$; $(1/2, 1/2)_0$

$^3P_2$; $(1/2, 3/2)_2$

$^3P_0$; $(3/2, 3/2)_0$

$^1D_2$; $(3/2, 3/2)_2$

$^3D_1$; $(3/2, 3/2)_1$

$^3D_2$; $(3/2, 1/2)_2$

A

C

F

B

D

E

$2p^5 3s$

$^1P_1$; $(1/2, 1/2)_1$

$^3P_1$; $(3/2, 1/2)_1$

Fig. 5. Six neon-like-ion laser transitions, so far observed for $Z \geq 29$. Levels are labeled in both LS (for Se $Z = 34$) and jj notations for convenience. Letters are used as labels to relate to Tables 1 and 2. The "B" and "C" transitions shown bold have produced the largest measured gain.

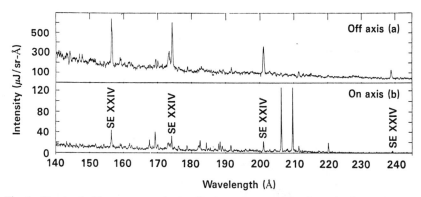

Fig. 6. Grazing-incidence spectra obtain off-axis or traverse (a) for the entire line, and on-axis (b) with a selenium target. The two strong Se^{24+} lasing lines are at 206 Å and 209 Å. The Se XXIV designations are for lines from Na-like Se^{23+} ions. (Adapted from Ref. 34.)

2.3.4. Measured Gain Coefficients

Convincing evidence for gain is demonstrated by a comparison between axial and transverse spectra, as shown in Fig. 6. In all of these LLNL and NRL experiments, the actual determination of a gain coefficient has been done exclusively by the exponential dependence of the axial emission (as discussed in Section 5.2 of Chapter 2). This method has proven to be the most convincing and accurate, even though many data points at various lengths are desirable. It also serves to identify anomalous effects with length, such as refractive losses (Section 2.4.5, Chapter 2). Measured gain coefficients are collected in Table 2. The accuracy here is expected[38] to be approximately 10%.

The largest gain has been measured on the two transitions shown bold in Fig. 5, namely "B" and "C". These originate from two different $J = 2$ upper states. The lower $J = 1$ states for these two transitions are also different, as indicated. Nevertheless, similar gain coefficients for Se^{24+} (see Table 2) of 4–5 cm^{-1}, measured over lengths of 1–2 cm, have resulted[46] in gain products as large as $GL = 17$ at a length $L = 4.4$ cm, before refraction limitations enter[42].

2.3.5. Output Power and Efficiency

Measured output powers of 2–5 MW have been reached[46] on 3p-3s transitions, with efficiencies (measured from the driver laser output) ranging from 10^{-6} for Se^{24+} to 10^{-9} for Y^{29+} ions.

Table 2

Measured Gain Coefficients [cm^{-1}] for Ne-Like 3p-3s Transitions.

Transition	A	B	C	D	E	F	Ref.
Species							
Cu^{19+}	2.0	1.7	1.7				38
Zn^{20+}	2.3	2.0	2.0				41
Ge^{22+}	3.1	4.1	4.1	2.7	4.1		39
As^{23+}			5.4				41
Se^{24+}	<1–2.7	4.0	3.8	2.3	3.5	≤1	36, 42, 47, 48
	2.6	4.9	4.9				41
Sr^{28+}		4.4	4.0				36
Y^{29+}		4–5	4–5				42
Mo^{32+}	0	4.1	4.2	2.9		2.2	35

2.3.6. The "J = 0–1 versus J = 2–1" Anomaly

As successful as have been the experiments on 3p-3s transitions in Ne-like ions, there remain some quite puzzling aspects. One of these concerns the particular pair of spectral lines ("B" and "C"—shown bold in Fig. 5) from $J = 2$ to $J = 1$ transitions which show the largest gain, and the $J = 0$ to $J = 1$ "A" line. One question is why the first two of these transitions, involving different levels, should consistently have the same gain coefficient for the many Z's tested.

Of equal or greater importance concerns the relative gains of "B" and "C" compared to that of "A". Atomic theory predicts that the electron collisional excitation rate into the $J = 0$ upper level for the "A" transition should be much larger than that into each $J = 2$ upper level[13]. All things considered, including collision/radiative mixing from other nearby levels and dielectronic recombination, the gain coefficient for the Se^{24+} $J = 0$ to 1 "A" line is expected[49] to be larger than that of either of the $J = 2$ to 1 "B" or "C" transitions by a factor of 1.5. This, however, is clearly not the case in the LLNL experiments, as described in the next section.

The experimental results are discussed next, followed by some proposed explanations. If it appears that an inordinate amount of emphasis is placed on these anomalies, it is felt that they may be of key consequence in scaling upwards to higher photon energy with improved efficiency. For example, in general gain coefficients appear at present to be limited to the $5–6$ cm^{-1} range, in spite of attempts to increase this for improved efficiency. The reasons for this apparent cap are not yet known. If there is a common limitation associated with the structure of the plasma medium, it could be of great importance to all schemes under study.

2.3.6a. THE LLNL RESULTS. In the initial thin-film Se experiments at LLNL only the $J = 2$ upper-level "B" and "C" lines showed measureable gain. No discernible emission was measured for the "A" transition originating on the $J = 0$ upper level[34,42]. This was extremely puzzling because, as mentioned above, modeling predicts that the "A" transition originating on the 3p $^1S_0$ term should have a larger gain coefficient. Data on higher-Z Ne-like ions from the same experiment generally supported such a disagreement. In later experiments with an extended gain length the "A" line was indeed observed. However, the intensity could not be compared directly with those of "B" and "C", because of the sensitivity required.

Reliable measurements of the gain coefficient for the "A" line also have been elusive. The available data have been interpreted at various times as giving gain coefficients varying[42,48] from $<1 \text{ cm}^{-1}$ up to 2.7 cm^{-1}. This spread also is puzzling because the data are usually sufficiently reliable to provide gain coefficients with an accuracy of $\pm 10\%$. (This includes two lines ("D", "E") of intensity similar to the "A" line.) The "A" gain coefficient deduced therefore has fluctuated over a range of approximately two to four or more times smaller than that for "B", rather than being larger as predicted. It also continues to be approximately three times smaller than theory predicts, whereas "B" and "C" are now within 50% agreement with theory.

2.3.6b. FURTHER DATA FROM NRL.   Further information on this apparent anomaly is provided by experiments[38-41] at the Naval Research Laboratory (NRL) described above. These involved most of the same transitions in Ne-like Cu, Zn, Ga, Ge, and As, as well as Se (see Table 1). In these experiments, the LLNL results for the "B" and "C" lines with a $J = 2$ upper levels were verified in that similar gain coefficients were measured (see Table 2, and also Fig. 26 of Chapter 2). The major difference in results was that $J = 0$ to 1 "A" gain lines were observed with approximately the same intensity and gain coefficient as for the two strong "B" and "C" $J = 2$ to $J = 1$ lines. Indeed, the "A" line for copper even showed a slightly larger gain coefficient than that for "B" or "C". It now appears[41] that the gain coefficient for the "A" line does not vary significantly with Z (see Table 2), but that the coefficients for the "B" and "C" lines increases with Z and reach a peak around arsenic or selenium. The latter trend also agrees with a recent Z-scaling analysis[10].

2.3.6c. OTHER SUPPORTING DATA.   Evidence that the observed anomaly may be associated with a collective medium phenomenon rather than a question of proper atomic modeling is provided by two recent experiments. In one[50], there are data on direct measurements of electron impact excitation cross-sections for exciting $n = 3$ levels in Ne-like Ba^{46+} in an electron beam ion trap ("EBIT") device that agree within 20% with theoretical predictions. In another experiment[32] at relatively low Z, spontaneous emission in thirty 3p-3s and 3d-3p transitions in Ne-like Ar^{8+} are measured in a well-diagnosed theta-pinch plasma at lower densities. The overall agreement with modeling, including the "A", "B" and "C" lines, indicates again that the transitions and rates included in the codes are adequate.

Hence it must be concluded that errors remaining in the atomic kinetics cannot explain alone the anomalies observed in the LLNL experiments. One important factor that needs to be measured in all of these experiments is the location of the gain region in the expanding laser-generated plasma. This is because various types of targets have been used successfully, ranging from ultra-thin foils to solid slabs.

2.3.7. *Possible Explanations for the J = 0, J = 2 Anomaly*

There have been numerous explanations put forth to explain these apparent anomalies. Some are described below.

2.3.7a. A LOWER $J = 0$ EXCITATION RATE.   This would be the easiest explanation, particularly since the upper "A" level is populated only from the ground state. However, the calculated excitation rates are considered to be very reliable. There are also the data described above from an electron beam ion trap and from a θ-pinch device that support the assumption that the calculated excitation rates are correct.

2.3.7b. COLLISIONAL MIXING.   One attempt at explaining the anomalously weak gain on the "A" transition compared to "B" or "C" has been to propose additional collisional redistribution. Strong collisional coupling between 3p levels could lead to localized equilibrium with a statistical distribution of populations according to $2J + 1$. This would produce an upper-state density ratio $N_u(J = 2)/N_u(J = 0) = 5$. Also, the corresponding 3p-3s transition probability ratio $A_{u\ell}(J = 2)/A_{u\ell}(J = 0) = 0.3$ and 0.6 for the Se^{24+} 206 and 209 Å $J = 2$ lines, respectively[13]. Hence, the product $N_u A_{u\ell}$ on which the gain coefficient depends is in the approximate ratio for $J = 2$ to $J = 0$ of 1.5–3, close to that observed for selenium. However, electron-collisional mixing rates are expected to be much less than radiative rates at the densities of interest[13]. This, along with the supporting experimental data on collisional rates and line intensities, mentioned above, argue against this hypothesis.

Ion-ion coupling collisions, notably strong for such transitions, have been considered also[51]. The calculated rate coefficients are plotted versus temperature in Fig. 7 for the strongest quadrupole de-excitation transition, namely, $3p^1S_0$-$3p^3D_2$. Accurate cross-sections for ion-ion collisions are difficult to obtain for such low-energy transitions, compared to electron-ion collisions. This is because the ion velocities are close to threshold, whereas electron velocities greatly exceeding threshold. Except for proton impacts above $kT = 400$ eV, electron-collisional deexcitation dominates in the $kT = 700–1000$ eV range of interest. Protons are not expected to be plentiful in the selenium-foil experiment; also selenium ion-ion collisional mixing is ruled out.

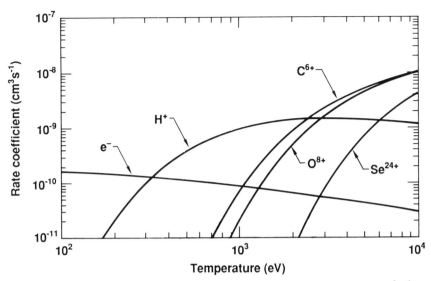

Fig. 7. Maxwellian-averaged rate coefficients for the deexcitation transition $2p^53p^1S_0$-$2p^53p^3D_2$ due to impact by electrons and by various ions. (From Ref. 51.)

2.3.7c. ALTERNATE PUMPING BY RECOMBINATION.   There have been numerous attempts to explain this gain anomaly by alternate pumping mechanisms (rather than electron collisional excitation from the Ne-like ground state) that might favor the higher J values of the 3p configuration. One such attempt attributes the pumping solely to electron-collisional recombination (see Sections 1 and 2 of Chapter 4) from the fluorine-like ion[18]. This suggestion was inspired in part by the general observation that F-like ions are present when amplification is strong[13,38]. However, this pumping process requires a very low temperature, which would normally occur late in time after the plasma has expanded and cooled. This is not consistent with time resolved measurements of the selenium laser lines, which show a maximum x-ray amplification at the peak of the driving (and heating) pulse[52].

2.3.7d. ENHANCED POPULATION BY DIELECTRONIC RECOMBINATION.   The inclusion of dielectronic recombination (see Section 5 of Chapter 4) in the modeling, a proeess which includes the promotion of a ground state electron and alters ionization balance, does indeed improve the agreement with the experiments. However, it falls short of a total explanation[49].

2.3.7e. ENHANCED PUMPING BY INNERSHELL IONIZATION.   Electron-collisional ionization of a 2s or 2p innershell electron from a $1s^22s^22p^63s$, p or d

Na-like ion to form an excited Ne-like ion has been shown to have a larger rate coefficient (within 20% accuracy) than for direct excitation from the neon-like ground state[53,54]. This effect is analyzed as an example of electron-induced ionization pumping in Section 1.2 of Chapter 5, and illustrated in Fig. 1 there. This process also favors high J values, and therefore could explain the higher gain for transitions originating on $J = 2$ sublevels.

However, most modeling of ionization balance predicts a low concentration of Na-like ions compared to Ne-like (and some F-like) ions under gain conditions. This is particularly true for a steady-state model. This tends to cancel the advantage of a large cross-section. The effect is illustrated in Fig. 8 from Ref. 53. It remains possible that there could be a very transient non-steady state ion distribution as a result of the temperature rising rapidly and ionization lagging. This would be particularly possible in the short pulse (0.4 ns fwhm) LLNL experiments. Then the Na-like density could be high and innershell ionization could favor the $J = 2$ upper levels, as is observed.

By the same token, slower rising 2 ns fwhm pulses such as present in the NRL experiments, may approach steady-state conditions favoring Ne-like

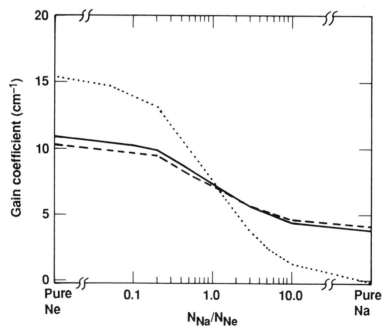

Fig. 8. Gain coefficient for 3p-3s transitions in neonlike selenium vs. ratio of sodiumlike to neonlike ions. Dotted line: $J = 0$ to $J = 1$ transition at 182.4 Å. Solid line: $J = 2$ to $J = 1$ transition at 209.6 Å. Dashed line: $J = 2$ to $J = 1$ transition at 206.3 Å. (From Ref. 53.)

ions and direct excitation pumping from the ground state. This would tend to favor $J = 0$ upper states, also as observed. Clearly, ionization balance measurements are needed to improve the knowledge of this possibility. The absence of lasing on other $2p^n$ sequences (e.g., F-like) can also be understood by innershell ionization pumping.

2.3.7f. PLASMA TURBULENCE. It is quite possible, given all of the evidence, that the gain distribution between $J = 2–1$ and $J = 0–1$ transitions might be affected by collective plasma effects in the lasant, rather than entirely from kinetics or alternate pumping. A proposed[55] explanation along these lines has to do with plasma turbulence effects on the Doppler width of the lasing lines and hence on the gain (see Section 2.2.2 of Chapter 2). The argument follows three logical steps:

(a) The ion-ion collisional mean free path in the Se experiments is shorter than or about equal to the radiation wavelength. Hence, velocity-changing collisions lead to a collisionally narrowed Doppler lineshape. It then follows that the emitting ions can no longer be assumed to be free streaming. This effect reduces the Doppler width by an estimated factor of three to five times.

(b) Phase-changing or "dephasing" collisions lead to collisional broadening. This effectively cancels the enhanced narrowing in (a) except for the $J = 0$ to 1 transition, which remains collisionally narrowed because of phase-shift cancellations between the upper and lower levels.

(c) If the plasma is turbulent, then varying Doppler motional shifts detune the gain lines. Such hydrodynamic turbulence may be associated with imperfections in the target foils and/or inhomogeneities in the driving-laser-beam. These are possibly reinforced by self-focusing and filamentation of the beam. Such turbulence has a most severe effect on the output intensity from the narrower $J = 0$ to 1 line. Weaker turbulence in the copper and germanium NRL experiments would then explain the presence of the $J = 0$ to 1 line there. Unfortunately, such turbulence is on such a microscopic scale as to make it virtually immeasurable with present techniques.

Another proposed plasma-related explanation for the $J = 0$ anomaly is the suggestion[56] that the $J = 0–1$ line is in a superfluorescent mode for selenium, while the other transitions are in an ASE mode (see Section 1 of Chapter 2), again because of the line narrowing mentioned in point (b). Hence, the $J = 0–1$ line radiates in random directions from hundreds of small cooperative volumes of size $\approx 180 \, \mu m$, instead of unidirectionally. The argument continues that in the NRL experiments, the densities and temperatures for the Cu and Ge plasmas, along with the collisional narrowing, are all less. Hence, all lasing lines are in the ASE regime. The authors go on to reject the need for a turbulence hypothesis as in the preceding explanation.

2.3.7g. SUMMARY OF THE $J = 0$ ANOMALY. The cause for the apparent anomaly in some gain experiments is still not understood, as documented in a recent LLNL committee report[49]. However, confidence in the atomic theory included in the modeling has risen to the point that other phenomenological differences are now more suspect. Clearly, this matter continues to be actively debated. More detailed and localized measurements of plasma conditions in the gain region, as well as level populations as a function of time, are needed to resolve this question. The final answers could very well have significant bearing on future laser designs in this configuration and others.

2.3.8. The Lack of F-like Ion Lasing

Another anomaly that surfaced during these experiments is that other potential laser isoelectronic sequences, notably the F-like $2p^5$ (ground state) ion, do not show any evidence of amplification. (This may or may not be related to the $J = 0$ versus $J = 2$ question.) This apparent anomaly holds true even when the plasma is heated to the point where O-like-ion radiation is observed. This is quite surprising. One natural assumption concerning this anomaly, as well as that for other $2p^n$ sequences, is that it is associated with the extra 2p-vacancies present.

Hence, it has been suggested[57] that double-electron transitions such as $2s^2 2p^4 3p - 2s^1 2p^6$ (for F-like ions) involving promotion of a 2s electron to the 2p shell, may be serving to radiatively deplete the upper-laser state density and thereby destroy the population inversion. Measurements of the subsequent 2p-2s innershell transitions would shed light on this possibility.

Another explanation for the anomaly may be that electron-collisional $n = 2$ innershell ionization is indeed a dominant pumping mechanism; and that lasing is unique to the neon-like ion because the sodium-like initial ion has a $n = 3$ electron present. In F-like ions formed from neon-like by $n = 2$ ionization, the 3p upper-laser state would not be formed. The same is continued through the $2p^n$ configurations to the B-like sequence.

2.3.9. Gain Experiments on 4d-4p Transitions in Ni-like Ions

The general concepts established so successfully for Ne-like 3p-3s transitions are currently being extended to 4d-4p transitions in high-Z Ni-like ions for even shorter wavelengths. In Eu^{35+} (Ref. 58) and Yb^{42+} (Ref. 59) the 4d-4p wavelengths are in the 65–71 Å and 50–56 Å ranges, respectively, as listed in Table 3. A similar sublevel structure exists as for Ne-like ions. However,

Table 3

Wavelengths and Gain Coefficients for Ni-Like 4d-4p Lasing^a.

Transition	Wavelengths [Å]		Gain Coefficients [cm^{-1}]		Ref.
	X^b	Y^c	X^b	Y^c	
Species					
Eu^{35+}	65.83	71.00	0.61	1.1	36, 58
Yb^{42+}	50.26	56.09	1.2	$(-)^d$	36, 59
Ta^{45+}	44.8^e				60
W^{46+}	43.1^e	49.3^e	5.5^e		60
Re^{47+}	41.6^e				60

^{a} All values are measured except where noted by superscript "e".
^{b} Transition X: $4d(\frac{3}{2}, \frac{3}{2})_0$-$4p(\frac{3}{2}, \frac{1}{2})_1$.
^{c} Transition Y: $4d(\frac{3}{2}, \frac{3}{2})_0$-$4p(\frac{5}{2}, \frac{3}{2})_1$.
^{d} Negative gain measured.
^{e} Calculated values.

in the Ni-like example only a $J = 0$ upper-laser state has shown any indication of gain (measured gain coefficients are included in Table 3). No positive gain is measured for the $J = 2$ upper levels, contrary to predictions[14]. The overall measured gain is marginal and is considerably lower for Ni-like ions than was the case for Ne-like ions. This is probably associated with the complexity and multiplicity of the level structure for the Ni-like sequence.

2.3.10. Alternate Drivers and Plasmas

There is no reason to expect that the pumping of $\Delta n = 0$ lasing transitions should be limited to laser-produced plasmas. Indeed, with a somewhat reduced level of radiative trapping affecting the lower-laser level population and therefore larger plasma dimensions (see Section 2.4.3 of Chapter 2), elongated pinch plasmas generated by pulsed power discharges deserve consideration as alternate plasma sources for x-ray laser experiments.

A concerted experimental effort was launched to demonstrate lasing in the 140–190 Å wavelength region in neon-like Kr^{26+} ions. These experiments were conducted at Physics International Corp. with the "PITHON" pulsed driver (see Section 4.2.2 of Chapter 2) and a gas puffed z-pinch load. In such a device, a jet of gas is injected into the vacuum diode region between two electrodes and the discharge takes place along the axis through this gas. In a sense, the gas replaces exploding wires used earlier, which had to be replaced on each shot and required vacuum release. In any case, these experiments produced no evidence of amplification, after several years of effort.

This failure illustrates a generic difficulty with z-pinch plasmas, namely the inherent axial non-uniformity and non-straightness associated with instabilities. Many attempts have been made to improve this, with marginal success so far (see Section 2.4.5 of Chapter 2 and Fig. 16 there). The increased output power and efficiency promised by success with such devices continues to provide the impetus for pursuing this approach.

3. ANALYSIS OF PHOTOEXCITATION PUMPING

Population of a selected upper-laser state from the ground state by absorption of photons at the proper frequency is an appealing method of pumping for short wavelength lasers. It avoids the problem of a larger pumping rate for the lower-laser state and therefore the need for a metastable upper state. (The latter was the situation for broadband plasma-electron collisional pumping described in Sections 1 and 2.)

The photoexcitation process can be described by the equation

$$X_o^{i+} + h\nu \rightarrow X_u^{i+}, \tag{28}$$

along with the energy-level diagram in Fig. 9. These definitions are quite similar to those for electron collisional excitation in Section 1. Here the photon $h\nu$ replaces the electron e. This also is a self-replenishment scheme, because the initial and final states are the same.

Another advantage in this method is that the pumping rate and the specific upper laser level can be controlled by the photon flux and chosen wavelength, respectively. The photon source conceptually resembles the flash-

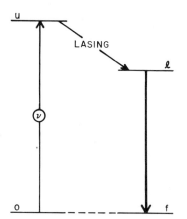

Fig. 9.   Energy level diagram for photoexcitation pumping.

lamp, familiar in long wavelength lasers. However, again, for ASE operation, gains of ≈ 100 times higher than in long-wavelength cavity operation require very high lasing-ion densities, and consequently high photon densities in the pumping source. Hence, while a narrow continuum source could be used for pumping, it is more efficient to use a bright line source which is frequency-matched to the absorption transition leading from the initial (o) to the upper (u) laser level[61,62]. This is referred to as photo-resonant or matched-line pumping between plasma ions: one dense plasma provides the photons in a spectral line; and another adjacent or congruent plasma with an appropriate absorption line in an ion serves as the more tenuous laser medium. The challenge then is to operate these two plasmas in close proximity for maximum photon coupling without excessive interaction. Such strong interaction could, for example, lead to overheating of the lasing plasma by the pumping plasma.

An obvious choice of geometries is coaxial, where the dense photon source is on axis and surrounded by the tenuous lasant plasma, with near total photon collection. The reverse is also possible with up to $\frac{1}{2}$ of the total collection efficiency. Here the lasant plasma is confined to a small diameter, which also could help to reduce opacity effects (Section 2.4.3 of Chapter 2).

3.1. Hydrogenic-Ion Analysis

As was the case for electron-collisional excitation pumping in Section 1, we begin the analysis by deriving an expression for the photoexcitation pumping rate P_{pe} in terms of the lasing wavelength $\lambda_{u\ell}$. With this and the decay rates, we then derive the fractional population ratio N_u/N_o, N_u alone, and finally the gain. Here we use the hydrogenic ion as a convenient example, with quantum numbers of $n_o = 1$, $n_u = 3$, $n_\ell = 2$, oscillator strengths[6] of $f_{ou} = f_{13} = 0.08$, $f_{\ell u} = f_{23} = 0.64$, wavelengths scaled as $\lambda_{uo} = \lambda_{31} = 1025/Z^2$, $\lambda_{u\ell} = \lambda_{32} = 6562/Z^2$, and finally $\Delta v/v = \Delta\lambda/\lambda = 3 \times 10^{-4}$ again. We will also use $Z = 6$ (C^{5+} ions) and $\lambda_{u\ell} = 182$ Å (Balmer-α transition) as a specific example at a wavelength convenient for comparison with Section 1.

3.1.1. Photoexcitation Pumping

We begin the derivation[63] of the photoexcitation rate P_{pe}, in analogy to collisional pumping, with

$$P_{pe} = N_v \sigma_{pe} c, \tag{29}$$

for a photon density N_v and a peak absorption cross-section σ_{pe}.

3.1.1a. PHOTOEXCITATION CROSS SECTION. We can relate the peak cross-section first to the frequency-integrated product $\int \sigma_{pe}(v)\, dv$ and then to the oscillator strength f_{ou} by

$$(\sigma_{pe}) \, \Delta v \approx \int \sigma_{pe}(v)\, dv = \pi r_0 c f_{ou}, \tag{30}$$

for a Doppler-broadened absorption line with a Gaussian shape and a half-width Δv. For the hydrogenic parameters given above, this becomes

$$\sigma_{pe} = 2.4 \times 10^{-18} \, \lambda_{uo} \quad \text{cm}^2, \tag{31}$$

for λ_{uo} in Angstrom units. For $Z = 6$, $\lambda_{uo} = \lambda_{31} = 28$ Å, and we find $\sigma_{pe} = 7 \times 10^{-17}$ cm^2 and $\sigma_{pe} c = 2 \times 10^{-6}$ cm^3 sec^{-1}. This latter rate coefficient is 2×10^5 times larger than that for collisional excitation pumping [see Eq. (8)], since it is a resonance process. Notice also that the rates for both pumping processes scale approximately the same with Z or wavelength.

In this hydrogenic model, the oscillator strength f_{ou} readily converts to the transition probability A_{uo} through Eq. (11) of Chapter 2. Using that equation, the cross-section can be related to the transition probability as

$$\sigma_{pe} = \frac{\lambda_{ou}^3}{8\pi c} \frac{g_u}{g_o} \frac{1}{\Delta v/v} A_{uo}. \tag{32}$$

3.1.1b. PHOTON DENSITY. Now that we have a relation for σ_{pe} in terms of A_{uo} for use in Eq. (29), we next seek a value for N_v. When the pumping ion is radiating with peak line emission at the saturation (black-body) level for a "brightness" temperature T_B (i.e., at high opacity), the intensity is given by

$$I_B = \frac{2h v_{uo}^3}{c^2} \rho \quad [\text{ergs/sec-cm}^2\text{-Hz-ster}]. \tag{33}$$

Here

$$\rho = \left[\exp\left(\frac{h v_p}{k T_B}\right) - 1 \right]^{-1} \quad \text{photons/mode} \tag{34}$$

is the occupation index[64] (from photon statistical analysis). The photon density N_v for absorption over a solid angle of 4π by the lasant is obtained from

$$N_v = \frac{4\pi}{chv} \int I_B(v)\, dv \approx 8\pi \frac{\rho}{\lambda_{uo}^3} \frac{\Delta v}{v}, \tag{35}$$

by substituting from Eq. (33). For a hydrogenic Lyman-α pump line, $h v_p = 3Z^2$ Ry/4. Also, we can assume that $k T_B = Z^2$ Ry/3 to define T_B. These lead to $\rho = 0.1$ photons/mode, using Eq. (34). (This is larger than a value of $\rho \approx 0.01$ often used in modeling[65], which may be more realistic.) In any case, assuming $\rho = 0.1$, Eq. (35) results in

$$N_v = 7 \times 10^{11} Z^6 \quad \text{cm}^{-3}. \tag{36}$$

For $Z = 6$, this gives $N_v = 3 \times 10^{16}$ cm^{-3}. This is 10^4 times smaller than $(N_e)_{opt}$ derived above [following Eq. (14)] for electron-collisional pumping. This smaller pumping density partially cancels the much larger rate coefficient above, for overall pumping efficiency.

3.1.1c. PUMPING RATE.   Substituting Eqs. (32) and (35) into Eq. (29) for ρ_{pe} and N_v, respectively, gives the convenient relation

$$P_{pe} = \left(\frac{g_u}{g_o}\right) A_{uo} \rho. \tag{37}$$

Allowing[6] $A_{uo} = A_{31} = 5.6 \times 10^7 Z^4$ sec^{-1} and converting Z to laser wavelength λ_{ul} (in Angstrom units), this becomes

$$P_{pe} = \frac{2.4 \times 10^{15} \rho}{\lambda_{ul}^2} \quad \text{sec}^{-1}. \tag{38}$$

For $\rho = 0.1$ and $\lambda_{ul} = 182$ Å, we have $P_{pe} = 7 \times 10^9$ sec^{-1}, which is similar to that for electron-collisional pumping of selenium at 207 Å and an optimum electron density [see Eq. (16)].

3.1.2. The Upper-State Density and Gain

3.1.2a. DEGREE OF PUMPING.   In estimating the fractional population density N_u/N_o in the upper-laser state, the collisional deexcitation rate from the upper level $n = 3$ to the nearest level ($n = 4$) can enter into the calculation if it is sufficiently large [see Eq. (17)]. A value for this rate can be estimated from Eq. (2) for $\langle g \rangle = 0.2$ (see Fig. 2), $f_{34} = 0.84$ (Ref. 6), and $\Delta E_{34} = Z^2$ Ry$(\frac{1}{9} - \frac{1}{16}) = 0.05 \times Z^2$ Ry. Assuming that $kT = Z^2$ Ry/3 and $(N_e)_{opt} = 3 \times 10^{14} Z^7$ [Eq. (28) of Chapter 2], we derive a collisional mixing rate of

$$N_e C_{34} = 5 \times 10^8 Z^4 \quad \text{sec}^{-1}. \tag{39}$$

On the hand, the total radiative decay rate from level three is[6]

$$A_{31} + A_{32} = 1 \times 10^8 Z^4 \quad \text{sec}^{-1}. \tag{40}$$

Hence, the maximum collisional excitation rate is five times as large as the radiative decay rate, and therefore contributes significantly to the depletion of the upper-laser level.

It may not be necessary to tolerate this additional depopulation rate, if we have sufficient external pumping photon flux. (This flexibility is an advantage over collisional pumping by plasma electrons.) Hence, we will assume a more conservative value of electron density, reduced by a factor of 10 but with the same Z-scaling, i.e.,

$$N_e = 3 \times 10^{13} \, Z^7. \tag{41}$$

With this value for electron density, we can simplify the analysis by ignoring collisional deexcitation from the upper-laser level. The fractional population is then given by Eq. (19) of Chapter 2 as

$$\frac{N_u}{N_o} = \frac{P_{pe}}{A_{uo} + A_{ul}} \approx \frac{P_{pe}}{2A_{uo}} = \left(\frac{1}{2}\right)\frac{g_u}{g_o}\,\rho = 0.5, \tag{42}$$

ideally, substituting from Eq. (37) and assuming[6] $A_{uo} \approx A_{u\ell}$, $\rho = 0.1$, and $g_u/g_o = 9$. This appears to be an unrealistically large fractional pumping ratio. In an actual experiment, it will most likely be reduced by a lower source brightness (i.e., reduced ρ), by a frequency mismatch between the pump and the absorber, and by less than ideal photon coupling from the pump line to the absorbing line. The latter two of these will be discussed in detail below, following the gain coefficient estimate.

3.1.2b. INITIAL- AND UPPER-STATE DENSITIES.   To obtain an expression for the initial state density N_o, we assume first that $N_o = N_e/3Z$ and use N_e from Eq. (41). This results in

$$N_o = 1 \times 10^{13} \, Z^6 \quad \mathrm{cm}^{-3}. \tag{43}$$

This is equivalent to

$$N_o = \frac{3 \times 10^{24}}{\lambda_{u\ell}^3}, \tag{44}$$

for $\lambda_{u\ell}$ in Angstrom units.

Also, from Eq. (42), ideally

$$N_u = \frac{1 \times 10^{24}}{\lambda_{u\ell}^3} \quad \mathrm{cm}^{-3}, \tag{45}$$

which equals $2 \times 10^{17} \, \mathrm{cm}^{-3}$ for $\lambda_{u\ell} = \lambda_{32} = 182$ Å. This again is likely to be a somewhat unrealistically high value for $N_u = N_3$.

3.1.2c GAIN COEFFICIENT.   We are now in a position to derive the gain coefficient, using Eq. (12″) in Table 1 of Chapter 2. The result is

$$G_{pe} = \frac{(3 \times 10^6)}{\lambda_{ul}^2} \quad cm^{-1} \tag{46}$$

$$= 90 \quad cm^{-1} \quad for \ Z = 6 \ and \ \lambda_{u\ell} = 182 \ \text{Å}.$$

This is for an inversion factor of $F = \frac{1}{3}$.

Again, this is an ideal case and probably unrealistically optimistic. Nevertheless, it is an often quoted level of gain in numerical modeling.[65] For more realism, it is appropriate to reduce Eq. (46) for an inexact match between the position, widths and profiles of the pumping and absorbing lines, and for the photon collection efficiency (i.e., a solid angle less than 4π). The rationales for these two revisions are discussed in the following sections.

3.2. Some Practical Criteria (and Limitations) for Gain

As mentioned above, some of the apparent advantage over collisional pumping is unrealistic, because we assumed an exact match between the energies (wavelengths) and line profiles for the pumping and absorbing transitions. In practice the two lines may be slightly separated and therefore only partially overlapping. Also, the optically thick pumping line may be 10 or more times broader than the absorbing transition. In either case, it is possible that most of the emission will not be absorbed. In addition, we will probably encounter less efficient photon coupling in a practical experiment, whereas we assumed an ideal 4π steradians collection solid angle in the present analysis. These factors can readily consume most of the 12-times advantage.

There is also the problem of opacity on the lower-laser level discussed at length in Section 2.4.3 of Chapter 2. It conceivably could be reduced in this case with sufficient pumping to allow a small density N_o.

3.2.1. Wavelength Match between Pump and Laser Ions

We now consider in detail the first limitation mentioned above, namely the match in transition energy. The more precise the match in central wavelength and line shape, the higher the efficiency and gain. Also, a narrow laser line is most desirable for high gain (see Section 2 of Chapter 2).

Numerous possible matches can be found between emission and absorption transition energies from compilations of energy levels and wavelengths for the xuv and soft x-ray regions[66-69]. However, certain additional criteria must be invoked for laser candidates. Besides the overlap, strong transitions must be chosen for both the pump and the lasant. For the pump this is necessary to assure maximum (blackbody) saturation emission, and for the lasant for maximum absorption (pumping efficiency) and gain. For the latter reason, the laser transition must also be strong. Finally, a rapid lower-laser-state depletion is necessary to assure quasi-cw operation.

Additional losses can be reduced by avoiding lasant ions whose upper-laser states are such as to autoionize or undergo photoionization induced by the laser beam itself. Both of these processes would decrease the upper-state population and the gain, while the latter would also absorb the laser emission itself. This means that one should preferably choose intermediate levels for lasing. (See also the summary of criteria in Section 1.3 of Chapter 7.) Besides these atomic considerations, it is also desirable to have pumping and lasing plasmas of comparable charge and temperature, in order to maintain a close proximity for the two plasmas and a large solid angle of coupling.

The control variable here is the pumping (blackbody) brightness temperature T_B, as well as the photon coupling solid angle. Notice that the density of the photon source does not enter into the gain coefficient formula, as long as an optically thick blackbody is generated for the pumping line. In fact, if the line is sufficiently opaque, additional broadening occurs[62] which is proportional to $(\tau_c)^{1/2}$, where τ_c is the optical depth from Section 2.4.3 of Chapter 2. Then line matches not normally sufficiently close with simply Doppler line broadening may become possible candidates, again with an associated loss in coupling efficiency.

A number of promising coincidences are listed in Tables 4–8 below, reproduced in large part from Ref. 70. Many wavelengths and matches have been updated from a recent wavelength compilation[69]. In these tables the subscripts p, a, and ul refer to the pumping, absorption, and most promising laser transitions, respectively. The quantities Δn_p and Δn_a represent the changes in quantum state for the pumping and absorbing transitions, respectively. The parameter λ_p represents the interacting wavelength in Angstrom units (shown to only two significant figures for conciseness). Likewise, $\Delta\lambda$ is the wavelength difference between the pumping and absorption transitions. These are combined in the figure-of-merit fractional wavelength match $(\Delta\lambda/\lambda_p) \times 10^4$. This parameter would be required to be ≤ 3 for purely Doppler broadened lines at a normal (equilibrium) ion temperature, according to the criterion adopted earlier. This is indeed a first-order constraint, and values exceeding three should not be rejected immediately. For example, additional line broadening may increase the overlap. Also, when relative streaming velocities between the two plasmas are $\gtrsim 10^7$ cm/sec, Doppler shifts estimated from $\Delta v_D/v \sim v/c \gtrsim 3 \times 10^{-4}$ can contribute significantly to a proper match.

The lasing transitions indicated begin on the pumped level and are for the shortest-wavelength transition. Some lasing may also take place on lower-energy transitions out of the same upper state, or follow electron cascade. These additional possibilities are not included. For certain indicated ([a]) examples, absolute pumping intensities have been measured, as discussed below.

3.2.1a. HE-LIKE AND LI-LIKE PUMP IONS. Tables 4 and 5 are organized mostly around intense He-like and Li-like pumping lines. This is because in many dense plasmas the np-$1s$ and nd-$2p$ resonance lines of these respective species consistently dominate in intensity.

Clearly a most promising combination is the He-like sodium and neon combination, shown in the first row of Table 4. This particular example is an excellent test case in that both the pumping and absorbing transitions can be calculated to a high precision. Not surprisingly, it has been a favorite subject of experimental investigations (see Section 2).

Table 4

Line Matches for (Mostly) He-Like **Pump** Ions.

Pump	Δn_p	λ_p (Å)	$\Delta\lambda/\lambda_p$ $\times 10^4$	Absorber/Lasant	Δn_a	λ_{ul} (Å)	Ref.
He-Like:							
Pump $n = 1$:							
$Na^{9+ a}$	2p-1s	11	1.5	Ne^{8+} (He-)	1s-4p	58	61
Si^{12+}	2p-1s	6.6	19	Al^{11+} (He-)	1s-3p	44	61, 71, 72
Mg^{10+}	3p-1s	7.8	22	Na^{10+} (H-)	1s-5p	36	61
P^{13+}	3p-1s	4.9	59	Si^{13+} (H-)	1s-4p	25	61
S^{14+}	3p-1s	4.3	21	P^{14+} (H-)	1s-4p	20	61
Fe^{24+}	2p-1s	1.9	20	Cr^{22+} (He-)	1s-3p	12	72
Sr^{38+}	2p-1s	0.8	1.4	Br^{33+} (He-)	1s-3p	5.4	72
Pump $n = 2$:							
Be^{2+}	2p-1s	100	0.12	O^{6+} (He-)	2p-4d	390	73
$B^{3+ b}$	2p-1s	60	22	Ne^{8+} (He-)	2p-4d	230	73
$C^{4+ b}$	2p-1s	40	38	Al^{11+} (He-)	2p-4d	160	73
$N^{5+ b}$	2p-1s	29	78	Si^{12+} (He-)	2p-4d	120	73
Ne-Like:							
$S^{6+ a}$	3d-2p	61	1.3	Ne^{7+} (Li-)	2s-5p	200	70
$S^{6+ a}$	3s-2p	73	1.5	Al^{5+} (O-)	2p-4d	440	70

[a] Absolute pumping intensities measured (see Table 9).
[b] Source of 2p-4d transition not specified in Ref. 73.

The second set of matches in Table 4 is less desirable in that the initial state is an excited 2p level, rather than the highly populated ground state. This is similar to the second set of hydrogenic pumping examples included in Table 6. There it is also pointed out that such transitions are relevant to lower-laser-level detrapping as discussed in some length in Chapter 2, Section 2.4.4. The last three entries in this section of Table 4 are based on 2p-4d and $\lambda_{u\ell}$ wavelengths given in Ref. 73 but of unknown source, hence the reliability is unknown.

A Ne-like S^{6+} ion pumping a Li-like Ne^{7+} ion is included in Table 4 as another promising example of a closed-shell pumping ion. (Closed shell ions have long lifetimes in transient plasmas, which can be a decided advantage.)

In Table 5, a Na-like Al^{2+} pump ion example is listed because this is one particular example where fluorescence has been measured in the xuv region[74]. The last entry in Table 5 is Ga-like Mo^{11+} pumping Kr-like Mo^{6+}, for which a close wavelength match has been measured[75].

3.2.1b. HYDROGENIC (LYMAN-α) PUMP IONS.   The 2p-1s Lyman-α resonance line in hydrogenic ions of nuclear charge Z_p is the common pumping transition in (most of) Table 6. This is typically a very intense line in high density plasmas. One particularly promising coincidence is between hydrogenic potassium and chlorine shown in the fifth row. This has been demonstrated[61,63]

Table 5

Line Matches for (Mostly) Li-like **Pump** Ions.

Pump	Δn_p	λ_p (Å)	$\Delta\lambda/\lambda_p$ $\times 10^4$	Absorber/Lasant	Δn_a	λ_{ul} (Å)	Ref.
S^{13+}	3d-2p	32	9.3	C^{4+} (He-)	1s-6p	160	70
Si^{11+a}	3d-2p	44	6.6	Mg^{9+} (Li-)	2s-4p	180	70
S^{13+}	3d-2p	32	0.90	Fe^{15+} (Na-)	3p-7d	84	70
P^{12+}	3d-2p	37	16	Mg^{9+} (Li-)	2s-6p	110	70
Cl^{14+}	3d-2p	28	18	Ni^{17+} (Na-)	3p-6d	66	70
Cl^{14+}	3d-2p	28	16	C^{5+} (H-)	1s-6p	110	76
Al^{10+a}	3p-2s	48	0.41	Mg^{8+} (Be-)	2s-4p	200	70
Si^{11+a}	3p-2s	41	11	B^{4+} (H-)	1s-3p	260	70
P^{12+}	3p-2s	35	18	C^{4+} (He-)	1s-3p	250	70
K^{16+}	3p-2s	21	6.7	N^{6+} (H-)	1s-3p	130	70
Cr^{21+}	3p-2s	13	9.5	F^{8+} (H-)	1s-3p	81	70
Al^{2+} (Na-)	5p-3s	560	0.06	C^+ (B-)	2p-5d	2100	74
Mo^{11+} (Ga-)	5s-4p	14	0.56	Mo^{6+} (Kr-)	4p-6s	600	75

[a] Absolute pumping intensities measured (see Table 9).

Table 6

Line Matches for H-like **Pump** Ions.

Pump	Δn_p	λ_p (Å)	$\Delta\lambda/\lambda_p$ × 10^4	Absorber/Lasant	Δn_a	λ_{ul} (Å)	Ref.
Pump $n = 1$ (mostly):							
O^{7+}	2p-1s	19	1.1	N^{6+} (H-)	1s-7p	81	70
Mg^{11+}	2p-1s	8.4	40	Na^{10+} (H-)	2s-4p	54	70
Al^{12+}	3p-1s	6.0	10	Sr^{28+} (Ne-)	2p-3d	160	77
P^{14+}	4p-1s	4.3	8.1	S^{14+} (He-)	1s-4p	28	61
K^{18+}	2p-1s	3.3	3.0	Cl^{16+} (H-)	1s-4p	17	61, 63
Mn^{24+}	2p-1s	1.9	4.0	V^{21+} (He-)	1s-4p	9.8	72
Ge^{31+}	2p-1s	1.2	0.28	Zn^{28+} (He-)	1s-3p	7.5	72
Pd^{45+}	2p-1s	0.55	1.6	Mo^{40+} (He-)	1s-4p		72
Pump $n = 2$:							
He^+	2p-1s	300	0.8	Be^{3+} (H-)	2p-4d	1170	63
Li^{2+}	2p-1s	130	0.44	C^{5+} (H-)	2p-4d	520	63, 73
Be^{3+}	2p-1s	76	0.40	O^{7+} (H-)	2p-4d	290	63, 73
B^{4+} [a]	2p-1s	49	0.41	Ne^{8+} (H-)	2p-4d	190	63, 73
C^{5+} [a]	2p-1s	34	0.30	Mg^{11+} (H-)	2p-4d	130	63, 73
N^{6+}	2p-1s	25	0.40	Si^{13+} (H-)	2p-4d	95	63, 73
Al^{12+}	2p-1s	7.2	7.0	Fe^{25+} (H-)	2p-4d	28	63
Ar^{17+}	2p-1s	3.8	?	K^{35+} (H-)	2p-4d	14	63

[a] Absolute pumping intensities measured (see Table 9).

by analysis to be very promising and convenient (similar Zs), if pumping plasmas of sufficient temperature ($kT \sim 2000$ eV) and density ($N_e \sim 10^{20}$ cm^{-3}) can be created.

The second group of matches in Table 6 occur with $n = 2$ to $n = 4$ Balmer-β absorbing transitions for ions in which the nuclear charge is $2Z_p$. This form of pumping can be better understood from the energy-level diagram in Fig. 10 of Chapter 2. (There, photoexcitation out of $n = 2$ and into $n = 4$ is considered for detrapping of a $n = 3$ to $n = 2$ Balmer-α laser scheme.) Following pumping, lasing would take place from $n = 4$ to $n = 3$ or perhaps even $n = 2$. Also shown in the same figure is $n = 2$ to $n = 6$ photo excitation by a Lyman-β photon. The matching wavelengths are determined essentially by the Rydberg formula and are naturally close. However, the plasma conditions are quite different for the different Zs involved. Also, the 2p absorbing state must be highly populated. This could require high opacity on the $n = 2$ to $n = 1$ transition in the absorbing ion. This could be contradictory to maintaining a low population density for an $n = 2$ or $n = 3$ lower-laser state. Hence, a careful choice of conditions is necessary for a successful experiment[63].

Pumping by the lithium-carbon as well as by the carbon-magnesium combinations shown in the 10th and 13th rows in Table 6, respectively, has been analyzed[78] in some detail. This analysis was directed primarily towards the possibility of detrapping the $n = 3$ to $n = 2$ Balmer-α lasing transition. Such pumpout of $n = 2$ population density can also lead to enhanced $n = 3$–2 population-inversion pumping by cascade from $n = 4$ to $n = 3$. These possibilities are discussed in Section 2.4.4 of Chapter 2, where quantitative pumping requirements are derived based on the above equations.

3.2.1c. MATCHES FOR F-LIKE LASER IONS.   In Table 7 are listed combinations of ions in which the lasant element is fluorine in either the He-like or hydrogenic state. These matches are unique in that, after initial suggestion[65] suitable matches were investigated through precise spectroscopic measurements[79] in laser-produced plasmas where shifts due to high density plasma effects were inherently included. Such small shifts would not be apparent in low-density experiments. The figures of merit ($\Delta\lambda/\lambda_p$) shown in Table 7 are therefore obtained under relevant experimental conditions, in contrast to those in the other tables which are based on compiled data from various sources.

3.2.1d. MATCHES FOR BE-LIKE LASER IONS.   Collected in Table 8 are five Be-like four-electron lasant ions. For these species, absorption takes place on the $2s^2\ {}^1S$-$2s4p\ {}^1P$ transition, with lasing following on the $2s4p\ {}^1P$-$2s3d\ {}^1D$ transition[81]. Lasing may also occur on the corresponding triplet transition

Table 7

Measured Line Matches for Fluorine Lasant Ions.

Pump	Δn_p	λ_p (Å)	$\Delta\lambda/\lambda_p$ × 10^4	Absorber/Lasant	Δn_a	λ_{ul} (Å)	Ref.
Fe^{17+}			4				
Cr^{19+}	3d-2p	14	0	F (He-)	1s-3p	99	65, 79, 80
Mn^{18+}			0				
Cr^{20+}	3d-2p		10				
Fe^{18+}	3s-2p		5				
		14		F (He-)	1s-4p	70	65, 79
Mn^{19+}	3d-2p		5				
Ni^{18+}	3s-2p		9				
Cr^{20+}	3d-2p						
		13	3	F (He-)	1s-5p	67	65, 79
Ni^{19+}	3d-2p						
Ni^{18+}			9				65, 79
Mn^{21+}	3d-2p	12	0.3	F (H-)	1s-3p	81	65, 79, 80
Cr^{21+}			10				65, 79

Table 8

Line Matches for Be-like Lasant Ions Pumped 2s to 4p.

Pump	Δn_p	λ_p (Å)	$\Delta\lambda/\lambda_p \times 10^4$	Lasant	λ_{ul} (Å)	Ref.
Mn^{5+} (Ca-)	4p-3d	310	0.33	C^{2+}	2163	81
P^{8+} (N-)	2p-2s	197	2.0	N^{3+}	1079	81
Al^{4+} (F-)	4s-2p	99	0.30	F^{5+}	513	81
Al^{7+} (C-)	3s-2p	76	1.7	Ne^{6+}	360	81
Al^{8+} (B-)	3d-2p	60	0.34	Na^{7+}	285	81
Al^{10+} (Li-)[a]	3p-2s	48	0.41	Mg^{8+}	230	81

[a] Table 5 also.

and possibly on 4f-3d transitions, through electron-collisional transfer of population (which also serves to decrease the gain on any particular transition). Quasi-cw operation is assured through rapid 3d decay to 3p and 2p levels. All of the six pumping transitions found are resonance transitions to a ground state, and hence are expected to be intense and optically thick when the particular ionic species is abundant.

3.2.2. Photon Coupling

We next consider the second practical limitation mentioned earlier, namely the efficiency of photon transfer from the pump plasma to the absorbing and lasing plasma. We consider three configurations.

3.2.2a. CONGRUENT PLASMAS.   We can start by relating the photon density in the laser medium to the power emitted from a completely and spherically congruent pumping source[70,82]

$$N_v = \frac{W_p}{h\nu_p} \frac{1}{V} \frac{r}{c} \approx 3 \frac{W_p}{h\nu_p} \frac{1}{4\pi r^2 c} \quad \text{cm}^{-3}. \qquad (47)$$

This is written for a spherical volume V determined by a radius r and a volumetrically congruent pumping plasma of output power W_p. In this ideal situation, the pumping power is emitted and absorbed over 4π steradians. Here, $h\nu_p$ is the energy of the pumping photon.

3.2.2b. SEPARATED POINT PLASMAS.   For achieving high gain, the important crucial variables here are high pumping power and small size. Realistically, the pumping and absorbing plasmas will be sufficiently different in character that they necessarily will be separated by some distance x to prevent undesirable interactions (e.g., overheating of the lasant, etc.). If a_L is the irradiated area of the lasing plasma, the relation between the photon density and the pumping power becomes, from Eq. (47),

$$N_v = \frac{W_p}{hv_p} \frac{\Omega}{4\pi} \frac{1}{a_L} \frac{1}{c} = \frac{W_p}{hv_p} \frac{1}{4\pi x^2 c} \quad \text{cm}^{-3}, \tag{48}$$

for a solid angle $\Omega = a_L/x^2$ (or for $\Omega = 4\pi$ and $a = 4\pi x^2$ in a spherical shell of radius x). Comparing Eqs. (47) and (48) shows a factor-of-three advantage for the congruent case, where $x \approx r$.

3.2.2c. SEPARATE ELONGATED OR COAXIAL PLASMAS.   For the important case of elongated rod-like pumping and lasing plasmas, both of length L and separated by a distance x, the result of integrating the line density W_p/L over the length of the pumping plasma is

$$N_v = \frac{W_p}{hv_p} \frac{1}{4\pi c} \frac{\eta}{xL}. \tag{49}$$

Here η is a geometrical factor of order unity. For $L = 10x$, η ranges from ~ 3 near the center of the rod to ~ 1.5 at the ends. The same result is obtained for a coaxial configuration in which the pump plasma is the core. In fact, the form of Eq. (49) remains valid for a number of other geometries, including the spherical shell case as well as the congruent plasma cases, using appropriate values for $\eta = 1$–3 and L a characteristic size for the plasmas[82].

3.2.3. Pump Requirements

3.2.3a. PUMPING POWER.   We can derive an expression for the power W_p required to achieve a particular gain product GL by first inverting Eq. (49) and assuming that $hv_p = hv_{ou}$. This gives

$$W_p = \frac{4\pi hv_p cx}{\eta} LN_v. \tag{50}$$

We can express N_v here in terms of the gain product GL by first converting N_v into N_u, using Eq. (29) along with the relation $N_u/N_o = P_{pe}/2A_{uo}$. Here we have assumed that the total decay rate $\sum A_u = 2A_{uo}$. Next we replace A_{uo} by σ_{ou} through Eq. (32) and N_u by $GL/\sigma_{u\ell}FL$ from Eq. (8) of Chapter 2. The result of these substitutions is (using $\sigma_{u\ell}/g_\ell = \sigma_{\ell u}/g_u$)

$$\frac{W_p}{xr_a} = \frac{64\pi^2 hc^2}{\eta F} \left[\frac{1}{\lambda_{ou}^4} \left(\frac{\lambda_{o\ell}}{\lambda_{\ell u}} \frac{g_o}{g_\ell} \frac{f_{o\ell}}{f_{\ell u}} \right) \frac{\Delta v}{v} \frac{GL}{N_o \sigma_{o\ell} r_a} \right]. \tag{51}$$

Here r_a is a characteristic absorption dimension (generally the radius) of the lasant plasma. Also, we have used $\sigma_{abs} \propto f_{abs}\lambda$ for the cross-sections [see Eq. (30)]. To be noted especially in this equation is the very strong inverse dependence on the pumping wavelength λ_{ou}. This wavelength therefore should be made as long as possible relative to that desired for lasing. Both short laser wavelengths ($\lambda_{u\ell}$) as well as weak ($g_1 f_{\ell u}$ low) laser transitions require increased pumping power, as is to be expected. Also to be noted in Eq. (51) is the absence of a dependence on the length L if we specify the gain product GL.

3.2.3b. PLASMA SEPARATION. Equation (51) has been used[83] to deduce mean distances $\langle x \rangle = (xr_a)^{1/2}$ from measured pumping powers on certain transitions indicated by a's in Tables 4, 5 and 6, assuming Doppler-broadened lines. For this it was assumed that $N_o \sigma_{o\ell} r_a$ and GL were both equal to five, so that the last factor in Eq. (51) is unity. Also a value for the inversion parameter of $F = 1$ was assumed, for an optimum degree of pumping, and $\eta = 3$. The results are reproduced in Table 9, where Δn_p represents the pumping transition. For most of the transitions measured, the peak intensities were already within a factor of three of the saturated blackbody level. The deduced dimensions of 20–100 μm are typical of plasmas produced by tightly focused laser beams. It was also found experimentally that the flux at the surface or even imbedded in the target crater itself was ~ 20 times larger (25 MW, see Ref. 82), which would yield higher gain or larger dimensions or shorter wavelengths, according to the scaling relations above.

Table 9

Pumping Power Calculated ($W_p/\langle x \rangle^2$) and Measured (W_p) and Characteristic Distance $\langle x \rangle$.

Pump	Δn_p	$\lambda u1$ (Å)	$10^{-10} (W_p/\langle x \rangle^2)$ (W/cm^2)	$10^{-4} W_p$ (W)	$\langle x \rangle$ (μm)
B^{4+}	2p-1s	190	0.026	3.9	120
C^{5+}	2p-1s	130	0.13	6.5	71
Al^{10+}	3p-2s	200	2.7	2.0	8.6
Si^{11+}	3p-2s	260	2.0	1.3	8.1
Si^{11+}	3d-2p	180	2.0	2.6	11
S^{6+}	3d-2p	200	4.1	0.32	2.8
S^{6+}	3s-2p	440	1.1	1.3	11
Na^{9+a}	2p-1s	11	31	120	20
				(2500)^b	(90)^b

^{a} For increased driving laser flux.
^{b} Measured in near-normal direction to the slab target.

Undoubtedly further systematic numerical comparisons and experiments will identify many more promising coincidences. Further experiments including line matches, fluorescence and gain will lead the way to what is perhaps the highest gains and efficiency, using this method of pumping with spectral-line "flashlamp" plasmas.

4. EXPERIMENTS WITH PHOTOEXCITATION PUMPING

There have been fewer successful experiments conducted with this approach than for those requiring single plasmas, such as described in Section 2. This is because of the added complexity when the pumping scheme requires two different plasmas which must be efficiently photon coupled but physically isolated. Nevertheless, such experiments are worth pursuing because of the potential for added control, high gain, and perhaps increased efficiency.

4.1. Vacuum-UV Experiments

The matched-line photoexcitation pumping approach, which typically involves $\Delta n = 1$ transitions on resonance lines, is quite amenable to the measurement of fluorescence and population inversions, preliminary to gain experiments. (This technique is described in Section 5.1 of Chapter 2.) Several examples of this are given next.

4.1.1. Al^{2+} Pumping C^+

There was a successful fluorescence experiment[74] involving C^+ lasing ions in an arc discharge as a potential lasant on a 5d-3p transition at 2138 Å, near the vacuum-uv region. Vuv pumping was accomplished here on a 2p-5d transition at 560 Å by photons from a Al^{2+} 5p-3s transition in a laser-produced plasma. This combination is listed in row 12 of Table 5.

4.1.2. Mn^{5+} Pumping C^{2+}

This match is listed in the first row of Table 8. Fluorescent enhancement by up to a factor of 150 was measured[84] on a 4p-3d line at 2177 Å in C^{2+} in the preceding experiment. Measured intensities from 4d-3p transitions also showed significant enhancements at 2163 Å and 1923 Å. Later, with an added cavity, a single-pass gain coefficient of 0.4 cm^{-1} was measured for the 2177 Å and 2163 Å lines[85].

4.1.3. C^{5+} Pumping Mg^{11+}

A xuv fluorescence experiment[86] designed to test the regular matches in hy-drogenic ions (rows >8 in Table 6) evolved from earlier successful Al^{12+} recombination-pumped fluorescence experiments performed at the Univer-sity of Rochester Laboratory for Laser Energetics (LLE). In this later version of the experiment, a hydrogenic Mg^{11+} plasma produced by a laser focused onto a solid slab expanded past a carbon strip from which a hydrogenic C^{5+} plasma was produced by the edge of the impinging laser beam. Lyman-α emission from the C^{5+} plasma was reported to have preferentially pumped $n = 2$ electrons from the Mg^{11+} ion into $n = 4$, as determined from enhanced resonance lines. The interpretation was unfortunately clouded by the simi-larity of the results to recombination pumping effects measured previously[87] without photoexcitation pumping (described in Section 2 of Chapter 4).

4.1.4. Na^{9+} Pumping Ne^{8+}

Of all the various combinations of pumping and lasant plasmas, the He-like Na/Ne combination is probably the most popular. This is primarily because of the exactness of and confidence in the pumping wavelength match near 11 Å (see Table 4, Row 1). The lasing wavelength is truly in the soft x-ray spectral region. While there is some preliminary fluorescent evidence for pumping, no gain experiments have been performed successfully so far.

4.1.4a. EARLY PUMP POWER MEASUREMENTS.   In an early experiment, the absolute power in the Na^{9+}-ion 2p-1s line near 11 Å was determined to be 25 MW, when heated by a 100 J Nd-glass laser[82,83]. It was also demon-strated[88] that a pure neon plasma could be generated from a solid layer of neon condensed onto a cryogenically cooled surface. The 25 MW of pumping power was marginal unless the plasmas were in very close ($\sim 100\ \mu m$) proximity.

4.1.4b. COATED/IMPLANTED-FOIL FLUORESCENCE EXPERIMENT.   A clever at-tempt to circumvent the plasma-interaction problem at such close distances was devised[89]. In an experiment, a thin aluminum plasma-isolating mem-brane (transparent to 11 Å radiation) was used as a target substrate. Neon atoms were implanted into the surface on one side and sodium was coated onto the opposite side. The target so made was irradiated by dual laser beams. Failure to generate fluorescence was attributed to a Doppler shift in the pumping radiation due to the relative streaming motion (velocity of $\Delta v_s \approx$

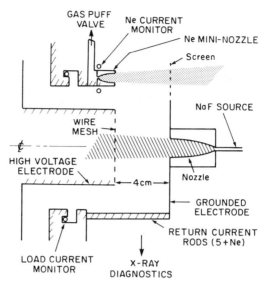

Fig. 10. Schematic of a gas filled z-pinch device that used photons from a sodium plasma to pump neon. The neon discharge replaces one of six return-current rods. (From Ref. 90, © 1988, IEEE.)

10^7 cm/sec) of the two plasmas. This experiment serves to point out the sensitivity of this scheme to the preciseness of the wavelength match.

4.1.4c. DUAL z-PINCH FLUORESCENCE EXPERIMENT.   Recently, there has been reported[90] some encouraging evidence of fluorescence at 230 Å from this Na/Ne combination in a pulse-power-driven z-pinch device (see Section 4.2.2b of Chapter 2). Vaporized sodium is compressed along the axis. The neon pinch is formed as part of the current return path. This experiment is shown schematically in Fig. 10. The amount of available Na^{8+} pump power at 11 Å was measured to be a very impressive 25 GW. This is 1000 times as large as that measured from the laser-heated NaF targets described earlier (Section 4.1.4a). A major problem again is straightness and uniformity of the lasant plasma over the gain length. This is aggravated in this particular experiment by distortion of the neon plasma by the Lorentz force from an oppositely directed current in the main (sodium) discharge.

4.2. A Doppler Shift Variation

It has been suggested[70] that Doppler streaming could actually induce or enhance a near resonance by causing a slight shift of frequency in the proper direction. This is the opposite effect to that attributed to the coated foil experiment described in Section 4.1.4b. However, at the present early stages of such experiments, concentration centers on natural (unshifted) matches.

REFERENCES

1. R. C. Elton, *Appl. Optics* **14**, 97 (1975).
2. L. J. Palumbo and R. C. Elton, *J. Opt. Soc. Am.* **67**, 480 (1977).
3. H. Van Regemorter, *Astrophys. J.* **136**, 912 (1962).
4. M. J. Seaton, in Atomic and Molecular Processes, D. R. Bates, ed. (Academic Press, New York, 1962).
5. R. C. Elton, "Atomic Processes", Chapt. 4 in "Methods of Experimental Physics, Plasma Physics, Vol. 9, H. R. Griem and R. H. Lovberg, eds. (Academic Press, New York, 1970).
6. W. L. Wiese, M. W. Smith and B. M. Glennon, "Atomic Transition Probabilities Vol. I: Hydrogen-Neon," NSRDS-NBS 4, 1966; and W. L. Wiese, M. W. Smith and B. M. Miles, Vol. II: Sodium-Calcium, NSRDS-NBS 22, 1969.
7. H.-J. Kunze, *Space Science Reviews* **13**, 565 (1972).
8. R. C. Elton, "Parameter regimes for x-ray lasing in plasmas," *Comments At. Mol. Phys.* **13**, 59 (1983). Because of printing errors, there is some confusion between "ell" and "one" in this paper. The author will provide a marked copy upon request.
9. A. N. Zherikhin, et al. *Sov. J. Quant. Electron.* **6**, 82 (1976); A. V. Vinogradov and V. N. Shylyaptsev, *Sov. J. Quant. Electron.* **13**, 303 and 1511 (1983), and references therein.
10. B. L. Whitten, R. A. London and R. S. Walling, *J. Opt. Soc. Am.* B **5**, 2537 (1988).
11. A. K. Bhatia, U. Feldman and J. F. Seely, *Atomic Data and Nuclear Data Tables* **32**, 435 (1985).
12. P. L. Hagelstein and R. K. Jung, *Atomic Data and Nuclear Data Tables* **37**, 121 (1987).
13. M. D. Rosen, et al., *Phys. Rev. Letters* **54**, 106 (1985).
14. S. Maxon, P. Hagelstein, B. MacGowan, R. London, M. Rosen, J. Scofield, S. Halhed and M. Chen, *Phys. Rev. A* **37**, 2227 (1988); also, W. H. Goldstein, J. Oreg, A. Zigler, A. Bar-Shalom, and M. Klapisch, *Phys. Rev. A* **38**, 1797 (1988).
15. R. C. Carmen and G. Chapline, Proc. International Conf. on Lasers '81 (STS Press, McLean, Virginia, 1982).
16. J. P. Apruzese and J. Davis, *Phys. Rev. A* **28**, 3686 (1983).
17. R. B. Miller, "Intense Charged Particle Beams," p. 19 (Plenum Press, New York, 1982).
18. J. P. Apruzese, J. Davis, M. Blaha, P. C. Kepple and V. L. Jacobs, *Phys. Rev. Letters* **55**, 1877 (1985).
19. B. L. Whitten, A. U. Hazi, M. H. Chen and P. L. Hagelstein, *Phys. Rev. A* **33**, 2171 (1986).
20. "Handbook of Laser Science and Technology, Vol. II, Gas Lasers", M. J. Weber, ed. (Chemical Rubber Publ. Co., Cleveland, Ohio, 1982).
21. J. D. Shipman, Jr., *Appl. Phys. Letters* **10**, 3 (1967).
22. R. W. Waynant, J. D. Shipman, Jr., R. C. Elton and A. W. Ali, *Appl. Phys. Letters* **17**, 383 (1970).
23. R. W. Waynant, J. D. Shipman, Jr., R. C. Elton and A. W. Ali, *Proc. IEEE* **59**, 679 (1971).
24. R. W. Waynant, *Phys. Rev. Letters* **28**, 533 (1972).
25. R. W. Waynant and R. C. Elton, *Proc. IEEE* **64**, 1059 (1976).
26. I. N. Knyazev, V. S. Letokhov and V. G. Movshev, *IEEE J. Quantum Electron.* **QE-11**, 805 (1975).
27. C. K. Rhodes, *IEEE J. Quant. Electron.* **QE-10**, 153 (1974).

28. H. A. Koehler, J. J. Ferderber, D. L. Redhead and P. J. Ebert, *Appl. Phys. Letters* **21**, 198, (1972); and *Phys. Rev. A* **9**, 768 (1974).
29. D. C. Lorents, in *Radiation Research* **59**, 438 (1974).
30. J. B. Marling and D. B. Lang, *Appl. Phys. Letters* **31**, 181 (1977).
31. C. Jupen, *Nuclear Inst. and Methods in Phys. Res. B* **31**, 166 (1988).
32. R. C. Elton, R. U. Datla, A. K. Bhatia, J. R. Roberts and H. R. Griem, *Bull. Am. Phys. Soc.* **33**, 1922 (1988); Phys. Rev. A, October 1989.
33. A. A. Ilyukin, G. V. Peregudov, E. N. Ragozin, I. I. Sobel'man and V. A. Chirkov, *JETP Lett.* **25**, 535 (1977).
34. D. L. Matthews, et al. *Phys. Rev. Letters* **54**, 110 (1985).
35. B. J. MacGowan, et al., *J. Appl. Phys.* **61**, 5243 (1987).
36. C. J. Keane, et al., *J. Phys. B* (October 1989).
37. D. L. Matthews and M. D. Rosen, *Scientific American*, p. 86, December 1988.
38. T. N. Lee, E. A. McLean and R. C. Elton, *Phys. Rev. Letters* **39**, 1185 (1987).
39. T. N. Lee, E. A. McLean and R. C. Elton, AIP Conf. Proc. No. 168, A. Hauer and A. L. Merts, eds., p. 125 (American Institute of Physics, New York, 1988).
40. R. C. Elton, T. N. Lee and E. A. McLean, *J. de Physique* **C9**, 359 (1987).
41. T. N. Lee, E. A. McLean, J. A. Stamper, H. R. Griem and C. K. Manka, *Bull. Am. Phys. Soc.* **33**, 1920 (1988).
42. D. L. Matthews, et al., *J. de Physique* **C6**, 1 (1986).
43. J. A. Cogordan and S. Lunell, *Physica Scripta* **33**, 406 (1986).
44. J.-P. Buchet, M.-C. Buchet-Poulizac, A. Denis, J. Desesquelles, M. Druetta, S. Martin and J.-F. Wyart, *J. Phys. B* **20**, 1709 (1987).
45. J. H. Scofield, LLNL, unpublished data; also M. J. Eckart, J. H. Scofield and A. U. Hazi, *J. de Physique* **Cl**, 361 (1988).
46. D. L. Matthews, et al., *Bull. Am. Phys. Soc.* **33**, 2041 (1988).
47. D. L. Matthews, et al., *J. Opt. Soc. Am. B* **4**, 575 (1987).
48. M. D. Rosen, et al., in "Atomic Processes in Plasmas," AIP Conf. Proc. No. 168, A. Hauer and A. L. Merts, eds., p. 102 (American Institute of Physics, New York, 1988); also R. A. London, M. D. Rosen, M. S. Maxon and D. C. Eder, *J. Phys. B*, (October 1989).
49. B. L. Whitten, et al., "Interim Report of the $J = 0$ Task Force," Report No. UCID-21152 (Lawrence Livermore National Laboratory, Livermore, California, 1987); also in Proc. Lasers '88 Conference, Lake Tahoe, Nevada, December 1988 (Soc. for Optical and Quantum Electronics).
50. R. E. Marrs, M. A. Levine, D. A. Knapp and J. R. Henderson, *Phys. Rev. Letters* **60**, 1715 (1988).
51. R. S. Walling and J. C. Weisheit, *Physics Reports* **162**, 1 (1988).
52. M. D. Rosen, et al., *Phys. Rev. A* **59**, 2283 (1987).
53. W. H. Goldstein, B. L. Whitten, A. U. Hazi and M. H. Chen, *Phys. Rev. A* **36**, 3607 (1987).
54. D. H. Sampson and H. Zhang, *Phys. Rev. A* **36**, 3590 (1987).
55. H. R. Griem, *Phys. Rev. A* **33**, 3580 (1986).
56. G. Hazak and A. Bar-shalom, *Phys. Rev. A* **38**, 1300 (1988).
57. R. C. Elton, T. N. Lee & W. A. Molander, *J. Opt. Soc. Am. B* **4**, 539 (1987).
58. B. J. MacGowan, et al., *Phys. Rev. Letters* **59**, 2157 (1987).
59. B. J. MacGowan, et al., *J. Opt. Soc. Am. B* **5**, 1858 (1988).
60. B. J. MacGowan, et al., *Bull. Am. Phys. Soc.* **33**, 2042 (1988); also S. Maxon, et al., *Phys. Rev. Letters* **63**, 236 (1989).
61. A. V. Vinogradov, I. I. Sobel'man and E. A. Yukov, *Sov. J. Quantum Electron.* **5**, 59 (1975).
62. B. A. Norton and N. J. Peacock, *J. Phys. B* **8**, 989 (1975).
63. R. C. Elton, "ARPA/NRL X-ray laser program," NRL Memorandum Report No. 3482 (Naval Research Laboratory, Washington, DC, 1977), pp. 92–114.

64. P. L. Hagelstein, "Physics of short wavelength laser design," LLNL Report UCRL-53100 (Lawrence Livermore National Laboratory, Livermore, California, 1981).
65. P. L. Hagelstein, in "Atomic Physics 9," R. S. Van Dyck and E. N. Fortson, eds., p. 382 (World Scientific Publ. Co., Singapore, 1984).
66. R. L. Kelly with L. J. Palumbo, Atomic and Ionic Emission Lines Below 2000 Å, Hydrogen through Krypton" NRL Report No. 7599 (Naval Research Laboratory, Washington, DC 1973).
67. S. O. Bashkin and J. O. Stoner, "Atomic Energy Levels and Grotrian Diagrams" (North Holland Publ. Co., New York, 1975, 1978) Vols. I through IV.
68. R. L. Kelly, Atomic and Ionic Emission Lines Below 2000 Å, Hydrogen through Argon," ORNL Report No. 5922 (Oak Ridge National Laboratory, Oak Ridge, Tennessee, 1982).
69. R. L. Kelly, *J. Phys. Chem. Ref. Data* **16,** Suppl. No. 1, Parts 1–3 (1987).
70. R. H. Dixon and R. C. Elton, *J. Opt. Soc. Am. B* **1,** 232 (1984).
71. J. P. Apruzese, J. Davis and K. G. Whitney, *J. Appl. Phys.* **53,** 4020 (1982).
72. W. E. Alley, G. Chapline, P. Kunasz and J. C. Weisheit, *J. Quant. Spectros. Radiat. Transfer* **27,** 257 (1982).
73. V. A. Bhagavatula, *J. Appl. Phys.* **47,** 4535 (1976).
74. J. Trebes and M. Krishnan, *Phys. Rev. Letters* **50,** 679 (1983).
75. U. Feldman and J. Reader, *J. Opt. Soc. Am. B* **6,** 264 (1989).
76. A. N. Zherikhin, K. N. Koshelev, P. G. Kryukov, V. S. Letokohov and S. V. Chekalin, *Sov. J. Quantum Electron.* **11,** 48 (1981).
77. P. Monier, C. Chenais-Popovics, J. P. Geindre and J. C. Gauthier, *Phys. Rev. A* **38,** 2508 (1988).
78. R. C. Elton, *Phys. Rev. A* **38,** 5426 (1988).
79. P. G. Burkhalter, G. Charatis and P. Rockett, *J. Appl. Phys.* **54,** 6138 (1983).
80. P. G. Burkhalter, D. A. Newman, C. J. Hailey, P. D. Rockett, G. Charatis, B. J. MacGowan and D. L. Matthews, *J. Opt. Soc. Am. B* **2,** 1894 (1985).
81. M. Krishnan and J. Trebes, *Appl. Phys. Letters* **45,** 189 (1984).
82. R. C. Elton, T. N. Lee and P. G. Burkhalter, *Nuclear Instruments and Methods in Physics Research* **B9,** 753 (1985).
83. R. C. Elton, T. N. Lee and W. A. Molander, *Phys. Rev. A* **33,** 2817 (1986).
84. N. Qi, H. Kilic and M. Krishnan, *Appl. Phys. Letters* **46,** 471 (1985).
85. N. Qi, H. Kilic and M. Krishnan, *J. de Physique*, Colloque C6, **47,** 141 (1986); also N. Qi and M. Krishnan, *Phys. Rev. Letters* **59,** 2051 (1987) and *Phys. Rev. A* **39,** 4651 (1989).
86. V. A. Bhagavatula, *IEEE J. Quantum. Electron.* **QE-16,** 603 (1980).
87. V. A. Bhagavatula and B. Yaakobi, *Optics. Comm.* **24,** 331 (1978).
88. R. H. Dixon, J. L. Ford, T. N. Lee and R. C. Elton, *Rev. Sci. Instrum.* **56,** 471 (1985); also NRL Memorandum Report 5534, April 30, 1985 for more detail.
89. M. H. Key, *Plasma Physics and Controlled Fusion* **26,** 1383 (1984).
90. S. J. Stephanakis, et al., *IEEE Trans. on Plasma Science* **16,** 472 (1988).

Chapter 4

Pumping by Electron Capture into Excited Ionic States

In this mode of pumping excited states in multi-ionized atoms, an electron is captured either from the continuum or from another atom. Conditions are such that there results a concentration in a particular state and a population density inversion.

1. ANALYSIS OF ELECTRON-COLLISIONAL RECOMBINATION PUMPING

One of the earliest suggestions for pumping of an x-ray laser involves free-electron capture into high bound quantum states (n) of an ion, followed by cascading downward to lower states[1]. The rate of radiative cascading downward from high-lying states is reduced by competing collisional mixing among states of higher energy. The net result is a population density inversion, under proper plasma conditions. An analysis for this approach follows.

1.1. Level Structure

Collisional recombination, sometimes referred to as "three-body" recombination, is the inverse process to electron-collisional ionization from excited levels. The combined recombination and cascade process is illustrated by the equation

$$X_0^{(i+1)+} + 2e \rightarrow X_n^{i+} + e \rightarrow X_u^{i+} + e, \tag{1}$$

and by the energy-level diagram shown in Fig. 1. The extra "spectator" electron results in an additional N_e dependence in the capture and pumping rates (see below).

146

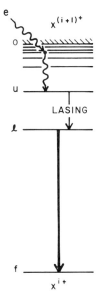

Fig. 1. Energy level diagram for pumping by electron-collisional recombination, followed by cascade, in a collisionally ionized plasma.

It is assumed that the ions exist plentifully in the initial state o, which is one stage of ionization higher than that of the lasing ion. In this case, such initial ions would be a product of quasi-equilibrium ionization balance created mostly by electron collisions in a plasma (in contrast to the photo-ionized scheme in Section 3). Notice that this is not a self-contained population-replenishment scheme, as was the case for excitation pumping discussed in the previous chapter. Hence, replenishment depends upon either reionization or on a flow of new ions into the lasant channel.

We can quantify the above concepts somewhat by defining a "collision limit" quantum level n'. For higher quantum states, collisional excitation is equally or more likely than radiative decay to take place. For quantum states lower than n', radiative decay dominates. Hence, a population density inversion can be generated between an upper level "u" above n' and a lower level "l" below n'. An alternate view is that levels u and higher are coupled to the next higher ion state, and that level l and lower are coupled to the ground state of the lasing ion.

1.2. Hydrogenic-Ion Analysis

The recombination pumping method lends itself readily to a simple one-electron hydrogenic analytical model for estimating gain and for isoelectronic extrapolation to short wavelengths. More detailed analyses are described in Refs. 2–4.

Here we follow an approximate procedure similar to that for photoexcitation in Section 3 of Chapter 3. The hydrogenic wavelength for any particular transition scales as Z^{-2} and hence the energy difference as Z^2. Also, we may assume that N_o, the initial-state (o) stripped-ion density, is given by $N_o = N_e/3Z$. Here Z is the nuclear charge of the stripped ion and the 3 represents the $\approx 1/3$ of the ions which are in the fully ionized initial state.

1.2.1. The Recombination Pumping Rate

The primary pumping process here, termed "collisional" or "three-body" recombination (labeled "cr"), involves the ion and two electrons. Hence, it is very sensitive to the particle density in the plasma. This is evidenced in an expression for the collisional-recombination pumping rate P_{cr} into a host of high-lying levels when written as[5,6]:

$$P_{cr} = (4\pi)^2 (n')^6 \alpha a_0^5 c \zeta^{-6} N_e^2 \left(\frac{\chi}{kT_e}\right)^2 \exp\left[\frac{\chi}{(n'+1)^2 kT_e}\right] \quad \text{cm}^{-3} \text{ sec}^{-1}, \quad (2)$$

or numerically as

$$P_{cr} = 1.4 \times 10^{-31} \zeta^{-6} N_e^2 (n')^6 \left(\frac{\chi}{kT_e}\right)^2 \exp\left[\frac{\chi}{(n'+1)^2 kT_e}\right] \quad \text{cm}^{-3} \text{ sec}^{-1}. \quad (3)$$

Here ζ is the effective nuclear charge seen by the recombining electron. In this case of an initially stripped atom, ζ is the nuclear charge Z. Furthermore, $\chi = Z^2$ Ry is the ionization potential, $\alpha = 1/137$ is the fine structure constant, and $a_0 = 0.53 \times 10^{-8}$ cm is the Bohr atomic radius. There are other formulations available. However, this one includes explicitly the collision limit and thus is particularly illustrative of the model.

The strong T_e^{-2} dependence in the above equations is very important, along with an effective N_e^3 scaling, for achieving a high pumping rate that exceeds radiative recombination (which leads preferentially into the lower-lying quantum states). This is illustrated by the ratio of collisional to radiative recombination rate coefficients R_{cr} and R_{rr}, respectively, plotted in Fig. 2. This is admittedly contradictory to the requirement (at least for thermal plasmas) of a high temperature to initially strip the atom of all electrons. Also, because of this strong dependence on density and temperature, the pumping is particularly sensitive to values of these variables. Hence, the proper operating conditions are in a quite limited region of plasma parameter space.

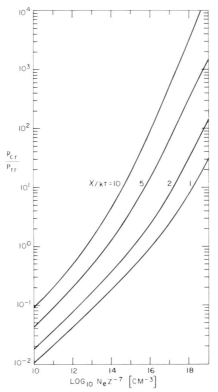

Fig. 2. The ratio of collisional (P_{cr}) to radiative (P_{rr}) recombination-pumping rate coefficients versus $N_e Z^{-7}$ for various values of χ/kT. (From Ref. 6.)

The collision-limit quantum number n' in Eqs. (2) and (3) is also somewhat plasma dependent and is given by[5,6]

$$n' = \frac{\zeta^{14/17}}{2^{2/17}\pi^{1/17}} \frac{(\alpha)^{6/17}}{(a_0)^{6/17}} \left(\frac{N_e}{\zeta^7}\right)^{-2/17} \left(\frac{\chi}{kT_e}\right)^{-1/17} \exp\left[\frac{4\chi}{17(n')^3 kT_e}\right] \quad (4)$$

or

$$n' = 1.26 \times 10^2 \left(\frac{N_e}{\zeta^7}\right)^{-2/17} \left(\frac{\chi}{kT_e}\right)^{-1/17} \exp\left[\frac{4\chi}{17(n')^3 kT_e}\right]. \quad (5)$$

This quantity is plotted in Fig. 3 and is seen to be mostly a function of the ratio of electron density N_e to effective charge ζ ($=Z$) (to the seventh power). This is the same Z-scaling as found in Chapter 2 and used in Chapter 3.

For the electron density N_e, we adopt a value intermediate between the very conservative value in Eq. (41) of Chapter 3 for photon pumping (i.e., $3 \times 10^{13} Z^7$ cm^{-3}) and the maximum value ($3 \times 10^{14} Z^7$ cm^{-3}) found in Eq.

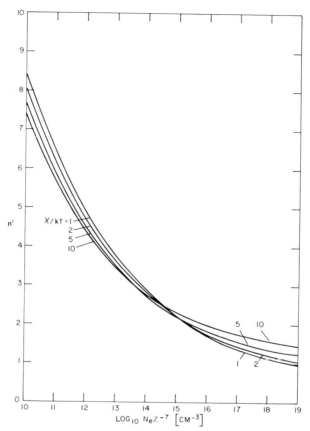

Fig. 3. The quantum number n' of the collision limit (i.e., the level from which radiative decay is about as probable as collisional excitation to higher levels) versus $N_e Z^{-7}$ for various values of χ/kT. (From Ref. 6.)

(28) of Chapter 2. Hence, we will use $N_e = 1 \times 10^{14} Z^7$ cm^{-3} and at the same time continue to ignore collisional depopulation to upper levels in arriving at a fractional inversion estimate. From Eqs. (4) and (5) and Fig. 3, this density implies that $n' = 2\text{–}3$ (2.5 mean). Assuming a (relatively cold) kT_e being a particular fraction of the ionization energy $\chi = Z^2$ Ry, namely, kT/Z^2 Ry $= 0.05$, a pumping rate P_{cr} can be derived from the above relations. It scales as Z^8 and is given by

$$P_{cr} \approx 700 \, Z^8 \quad \text{sec}^{-1}. \tag{6}$$

For $\lambda_{ul} = 6562/Z^2$ in Angstrom units, this becomes

$$P_{cr} \approx \frac{(1 \times 10^{18})}{\lambda_{u\ell}^4} \quad \text{sec}^{-1}. \tag{7}$$

Hence, P_{cr} scales upwards very rapidly with decreasing wavelength ($\sim \lambda_{u\ell}^{-4}$). As an example, $P_{cr} = 1 \times 10^9$ sec^{-1} for the currently popular C^{5+}-laser operating at $\lambda_{u\ell} = 182$ Å on the $n = 3$ to $n = 2$ Balmer-α transition.

1.2.2. The Upper-State Density and Gain

1.2.2a. THE DEGREE OF PUMPING.   From Eq. (19) of Chapter 2 and ignoring collisional depopulation from $n = 3$ to $n = 4$, we have for the ratio of upper- to initial-state densities

$$\frac{N_u}{N_o} \approx \frac{P_{cr}}{2A_{uo}}. \tag{8}$$

Next, using $A_{uo} = A_{31} = 5.6 \times 10^7 \, Z^4 = 2.4 \times 10^{15}/\lambda_{ul}^2$ sec^{-1} from Ref. 7, and Eq. (7) above, we have

$$\frac{N_u}{N_o} = \frac{200}{\lambda_{ul}^2} = 6 \times 10^{-3}, \tag{9}$$

for $\lambda_{ul} = \lambda_{32} = 182$ Å.

1.2.2b. INITIAL- AND UPPER-STATE DENSITIES.   Assuming as in Section 3 of Chapter 3 that

$$N_o = \frac{N_e}{3Z} = \frac{9 \times 10^{24}}{\lambda_{u\ell}^3} \quad \text{cm}^{-3}, \tag{10}$$

we have for the upper-state density

$$N_u = \frac{(2 \times 10^{27})}{\lambda_{u\ell}^5} \quad \text{cm}^{-3}. \tag{11}$$

This equals 1×10^{16} cm^{-3} for $\lambda_{u\ell} = 182$ Å.

1.2.2c. THE GAIN COEFFICIENT.   We can relate this upper density to a gain coefficient using Eq. (12″) given in Table 1 of Chapter 2. The result is

$$G_{cr} = \frac{(8 \times 10^9)}{\lambda_{u\ell}^4} = 7 \quad \text{cm}^{-1}, \tag{12}$$

for $\lambda_{u\ell} = 182$ Å again. Here we use a population inversion factor of $F = 1/3$, $f_{23} = 0.64$, and $g_\ell/g_u = 2/3$ from Ref. 7. This shows a strong $\lambda_{u\ell}^4$ dependence, which is very desirable for scaling to short wavelengths[8]. In actual experiments the gain coefficient is measured (see Section 2) to be in the range ~ 5–8 cm^{-1} which is very close to the above estimate.

1.2.3. Opacity Effects

We have not included any lower-level trapping effects (discussed in Section 2.4.3 of Chapter 2) in the above analysis. For hydrogenic-ion lasing, trapping can be a particularly severe problem because the resonance lines have such large radiative strengths. We can better understand the scaling of this problem at shorter wavelengths using Eq. (31) of Chapter 2, which is (for level "f" the same as level "o"):

$$\tau_c = 1.1 \times 10^{-16} f_{o\ell} \lambda_{\ell o} N_o r \left(\frac{\mu}{kT} \right)^{1/2}, \tag{13}$$

where kT is in eV. For hydrogenic C^{5+}, $Z = 6$, $\lambda_{\ell o} = 34$ Å, $f_{o\ell} = 0.42$, the atomic mass number $\mu = 2Z = 12$, and $kT = 24$ eV ($= 0.05 \times Z^2$ Ry again). At the electron density assumed above, namely $N_e = 1 \times 10^{14} Z^7$ ($= 3 \times 10^{19}$ cm^{-3} for carbon), for $N_o = N_e/3Z$ again, and assuming a plasma radius of $r = 25$ μm, we have $\tau_c = 3$. This is not sufficiently large to cause a significant $n = 2$ overpopulation.

However, for scaling to shorter wavelengths and higher densities, reducing the plasma radius drastically (~ 10 times) is not practical, while maintaining a straight plasma over lengths of mms to cms. Therefore, short of significant success with auxiliary detrapping methods (Chapter 2, Sections 2.4.3 and 2.4.4), the remaining option is to reduce the electron density. An added advantage of a decrease in electron density would be a reduction in refraction losses, as discussed in Section 2.4.5b of Chapter 2.

This implies that opacity on the relaxation resonance line serves to limit the plasma density, as we attempt to scale from $Z = 6$ upwards to, say, $Z \geq 13$ and $\lambda_{u\ell} \leq 39$ Å. We can evaluate this limitation and the associated effect on gain starting with Eq. (13). For this purpose, we set $f_{o\ell} = f_{12} = 0.42$, $\lambda_{\ell o} = \lambda_{21} = 1216/Z^2$, $N_o = N_e/3Z$, $\mu = 2Z$, and $T = 0.05 \times Z^2$ Ry again and require that $\tau_c = 1$ be maintained in the isoelectronic extrapolation. This results in a limiting electron density of

$$(N_e)_{\text{lim}} = \frac{3 \times 10^{17} Z^{3.5}}{r} \quad \text{cm}^{-3}, \tag{14}$$

for r in micrometers. This density scales with Z to a power one-half that assumed previously. If we now allow r to decrease linearly with Z, [i.e., $r = 25(6/Z)$ μm], we can assure that the gain coefficient G in Eq. (12) will continue to at least scale upwards as $Z^{0.5}$ for shorter wavelengths, using $(N_e)_{\text{lim}}$ instead of $N_e = 1 \times 10^{14} Z^7$ cm^{-3}. The result is a limiting electron density independent of r, i.e.,

$$(N_e)_{lim} = 2 \times 10^{15} \, Z^{4.5} \quad cm^{-3}. \tag{15}$$

This equals 7×10^{18} and $2 \times 10^{20} \, cm^{-3}$ for $Z = 6$ (C^{5+}) and $Z = 13$ (Al^{12+}), respectively. These are both reasonable for existing technology. Hence, with a tradeoff between practical size and density limitations versus gain scaling, recombination pumping remains promising for extrapolation to the $\lambda_{ul} = 40$ Å region.

If these limitations set by opacity on the relaxation line can be removed somehow by "detrapping" (see Sections 2.4.3 and 2.4.4 of Chapter 2), rapid scaling of gain with Z will be limited solely by the electron density and length achievable experimentally. Of course, any associated effects of additional line broadening (e.g., Stark effect) and refraction at increased density must still be confronted.

1.2.4. Temperature Reduction for High Gain

In achieving maximum gain with recombination pumping, the minimum electron temperature (and maximum pumping) during lasing is usually determined by how much cooling can be accomplished. This is after the plasma is first heated to $kT \approx Z^2 \, Ry/3$ to remove all of the electrons by collisional ionization. (An exception to this is photoionization of a cold plasma, discussed in Section 3.) To be fully effective, this cooling must be performed very rapidly, i.e., in a time shorter than or at least comparable to a characteristic time for producing the population inversion by capture and cascade. A value of $kT_e/\chi = 0.05$ for the temperature factor taken above is a reasonable goal. This was assumed intentionally to be somewhat less than a maximum ~ 0.1, above which electron-collisional excitation from the ground state into the lower-laser level rapidly depletes the population and gain[9] (see Figs. 17 and 18 of Chapter 2).

1.3. Extensions to Non-Hydrogenic Ions

A possible extension of the basic hydrogenic concept described above to more easily produce plasma ions with perhaps increased overall efficiency would appear to be desirable. One possibility is alkali-like ions with single outer electrons, for a corresponding increase in quantum numbers for the laser transition. For example, hydrogenic $n = 3$–2 lasing translates into 4–3 (or 5–3) in Li-like and 5–4 in Na-like ions[10]. The initial "o" states are He-like and Ne-like, respectively. Both of these are known to be highly populated species in transient plasmas. Care must be taken (with detailed modeling) to assure that the collision limit defined in Eqs. (4) and (5) is properly located between the laser levels.

2. RECOMBINATION EXPERIMENTS: COLLISIONALLY IONIZED PLASMAS

We discuss here experiments involving the recombination of ions existing in a quasi-equilibrium and quasi-neutral plasma, as analyzed in Section 1. This is perhaps the most systematically studied system. As such, the concept has been tested at visible wavelengths, fluorescence studies at low densities have been used to measure xuv population-inversion densities, and ASE has been demonstrated in the soft x-ray spectral region. The use of this method of pumping is also the most widespread, in terms of the number of laboratories involved. It has the advantage of relative simplicity inherent in a single-component plasma.

The general approaches that have been pursued experimentally are described in this section. The major results and references are compiled in Table 1. There measured wavelengths, population inversions (PI) and gain coefficients are tabulated. The latter were determined by many of the techniques described in Section 5 of Chapter 2. It can be noted in this table that many of these experimental results are quite recent. This indicates the current relatively high level of interest in this approach, in the quest to achieve lasing in the biological "water window" spectral region (see Chapter 7, Section 4.4.1). Some of the results are based on rather sparse data and some gains are marginal (< 1.5 cm^{-1}). As such, some results included remain preliminary and subject to verification.

2.1. Experimental Conditions

2.1.1. Two Approaches

There are basically two classes of recombination-pumped x-ray lasers that have been used successfully in achieving ASE. The most extensively studied involves a short-pulse laser-produced plasma of very small dimensions. The plasma is then encouraged to expand and to cool rapidly for enhanced recombination. The second class of experiments involves a long-pulsed laser-produced plasma confined in a magnetic field. Recombination occurs at sustained density as the plasma cools by radiation and conduction. These will each be discussed separately in Sections 2.2 and 2.3.

2.1.2. Desirable Ionic Species

Most experiments to date have been carried out with stripped initial atoms. Recombination and cascading then lead to preferential population in $n = 3$ or 4 levels of hydrogenic ions, as analyzed in Section 1. Population inversion and gain takes place on 4–3, 4–2 and, most popularly, 3–2 transitions. Lower-level depletion than occurs on Lyman-α or Lyman-β resonance transitions from $n = 2$ or $n = 3$ levels, respectively, to the ground state.

An alternative combination involves He-like initial ground-state ions recombining for lasing on Li-like ions. Lasing then occurs on 5–4 or 5–3 transitions, with depletion on 4–2 or 3–2 transitions, respectively. For the same atom, these wavelengths are longer than for hydrogenic ions. One advantage of Li-like lasing ions is the high density of initial ions that can be obtained with less power invested than for complete stripping. Another advantage may be reduced radiative trapping, compared to the strong Lyman depletion transitions in hydrogenic ions.

Experiments on both of these combinations will be described in the following sections. One would think that the successes with hydrogenic ions would carry over naturally to the quite similar He-like lasing ions. Yet that is not the case. While population inversion has been measured for He-like Al^{11+} (see Table 1), gain on the same transitions has not yet been observed. (The reader will recall that a similar apparent anomaly arose in Chapter 3, Section 2, where it is remarked that Ne-like lines are found to lase well but not F-like or lower series members for similar transitions.) Another possibility that has received virtually no experimental attention so far is Ne-like ions recombining for lasing in Na-like ions, mentioned in Section 1.

2.2. Short-Pulse-Heated, Expanding, Cooling, Laser Plasmas

In most successful gain experiments to date, recombination pumping follows a short burst of ionizing energy in the plasma. It is important that the heating and ionization be completed before the plasma has expanded freely to a significant degree. For example, a 10 μm fiber target (described below) requires a 100 ps heating pulse for a typical expansion velocity of 10^7 cm/sec. The highest concentration of power is available in focused laser beams.

Table 1

Measured Wavelengths and Gain Coefficients for Recombination Pumping

Species	Transition	Target[a]	Wavelength [Å]	Gain Coef.[b] [cm^{-1}]	Ref.
Hydrogenic:					
C^{5+}	3p-2p	S	182	PI	12–14
	3–2	Fi	182	25	15
		M		6.5	16[c]
		Fo		3	17
		Fi		4.1	18–20
		Fo		2.0	21
O^{7+}	3–2	Fo	102	≈0.5	22
	3–2	Fo	102	2.3	21
F^{8+}	3–2	Fi	81	5.5	19, 20
		Fo		2.0	21
He-Like:					
Al^{11+}	4-6p–3, 4p	S, Fo	76–280	PI	23, 24
Li-Like:					
O^{5+}	4f-3d	M	520	1.8	25
Ne^{7+}	4f-3d	M	292	PI	25
Al^{10+}	5d-3p	Fo	104	1.0	21
	4d, 5d-3d	S	106	PI	26
	5f-3d	S	106	2–2.5	27[c]
		Fi		3	19, 20
		Fo		1.5	21
		Fo		3.5	28
	4d-3p	Fo	151	4.5	28
	4d-3d	S	155	PI	26
	4f-3d	M	155	3–4	29
		Fo		4.1	28
Si^{11+}	4f-3d	M	129	1–2	29
S^{13+}	5f-3d	S	65	1	30
Cl^{14+}	4f-3d	Fi	80	3.5	19

[a] Targets: S = slab, M = magnetically confined, Fi = fiber, Fo = foil.
[b] "PI" designates population inversion measurements.
[c] Unusually large pumping efficiency (Section 1.2, Fig. 1, Chapter 7).

Hence, all such fast experiments have incorporated short-pulsed lasers as drivers.

After the initial heating and ionization to the proper stage, sudden cooling must be accomplished for enhanced recombination and gain while the lasing plasma is intact. The method of such cooling is a feature that often distinguishes a particular experiment. The most natural cooling process, at least for laser-produced plasmas, is adiabatic expansion[11]. Two-dimensional expansion would be the most desirable[2]. This is sometimes approached experimentally, but usually the expansion is somewhere between one- and two-dimensional.

2.2.1. Population Inversion Experiments

Prior to gain experiments at high density, quantitative fluorescence experiments were performed. This was possible because both the upper- and lower-laser states decay spontaneously, i.e., are not metastable. Hence, in all of these experiments population inversions were conveniently deduced from anomalous intensities in the resonance spectral series. This fluorescence technique is discussed at length in Section 5.1 of Chapter 2.

The results are included in Table 1 and designated by "PI" for population inversion. This work was performed mostly with modest point-focused long-pulse driving lasers, solid-slab targets, and at relatively low densities, where opacity was also not severe. Early experiments on hydrogenic ions such as C^{5+} were extended to He-like Al^{11+} ions with a larger laser (and with a novel cooling blade placed close to a slab target). The extension of population-inversion experiments continued to Li-like Ne^{7+} and Al^{10+} ions also, as indicated in Table 1.

2.2.2. Gain Experiments

Following these early inversion measurements, it was recognized that significant gain would occur only at much higher densities. The density required is so high in fact that radiative trapping on the lower-laser level can become a significant deterrent for very short wavelength operation and finite plasma dimensions. This trapping problem is discussed in Section 1. While there are suggestions that trapping can be alleviated somewhat by Doppler shifting between the emitting and absorbing ions in the laser channel (see Section 2.4.3b of Chapter 2), such an effect is difficult to quantify explicitly in experiments.

While maintaining a high density in the laser region, it is also very important that the temperature be kept low. As analyzed in Section 1, this is necessary both to enhance the pumping rate and gain and to prevent excitation-pumping of the lower-laser level.

An expanding and rapidly cooling laser-produced plasma can be quite naturally expected to produce these required conditions at some point in space and time. Fortunately, it has turned out to be possible to create the proper conditions without particularly exotic experimental intricacies.

2.2.2a. SMALL-FIBER- AND RIBBON-TARGETS.   In one type of experiments indicated by Fi in Table 1, very thin (2–10 μm in diameter) carbon fibers are vaporized by transverse irradiation from a short pulse line-focus driving laser. A layout for some early experiments is diagrammed in Fig. 27 of Chapter 2. This fiber-target design was chosen to give two-dimensional radial expansion for an optimum cooling rate (providing the fiber is illuminated circumferentially). An extension of the fiber-target design that is closer to a one-dimensional expansion and permits a broader focus for more efficient coupling of the driver beam is a narrow- and thin-ribbon target.

2.2.2b. THIN-FOIL TARGETS.   A variation on these approaches is to use thin foils as targets. There, two-dimensional expansion is approached to even a lesser degree than with ribbons. Such thin-foil experiments are indicated by "Fo" in Table 1. When the foil is ultra-thin ($\lesssim 1000$ Å), it may be possible to explode it with drivers from both sides and thereby simulate multi-dimensional expansion and cooling.

2.2.2c. SOLID SLAB TARGETS.   All of this is not to say that slab-target plasmas (designated by "S" in Table 1) will not serve also as useful x-ray laser media. As a reminder, the early measurements of population inversion described above were achieved with solid-slab targets. Significant gain coefficients for Li-like Al^{10+} are measured (see Table 1). Slab targets were also found to be practical for 3p-3s lasing, as described in Section 2 of Chapter 3. Indeed, in some thick ($\gtrsim 1$ μm) foil experiments with short-pulse lasers, surface ablation dominates the plasma production during the time of interest and the results are comparable to those of slab targets. Slab targets are of course limited to single-sided illumination.

2.3. Laser-Produced Plasmas Confined in a Solenoid

Recombination-pumped gain in hydrogenic carbon, as well as some Li-like ions, has also been observed in a distinctly different configuration. Here the plasma is produced by an axially directed long pulse (50–70 ns), long wavelength (10 μm), CO_2 laser. The fully ionized plasma is confined by a magnetic field, rather than allowed to expand freely. This is shown schematically in Fig. 4 for aluminum; the results for various elements are disignated by "M" in Table 1. The magnetic field presumably serves to sustain the electron density as the plasma cools by radial thermal conduction across the field lines. It has been found important to include a thin blade along the length of the

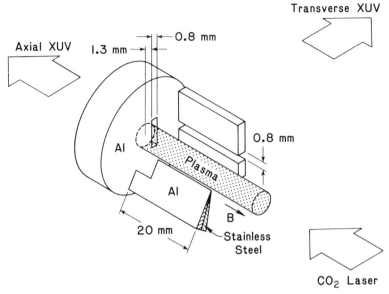

Fig. 4. An aluminum-disk target geometry with a vertical slot and an aluminum/stainless-steel composite blade attached perpendicular to the target surface. (From Ref. 29, © 1988 *IEEE*.)

plasma and adjusted to the very edge. This presumably stimulates the required cooling by conduction and enhanced radiation. Evidence for amplification again comes from a comparison of axial and transverse intensities, as shown in Fig. 5. Also, it is conceivable that there is a contribution to the population inversion from charge exchange between the C^{6+} ions and neutral atoms vaporized from the blade (see Sections 6 and 7 of this chapter).

Results[29] from numerical modeling indicate that the lasing takes place in a 100 μm-thick outer annulus at an overall diameter of 1.5 mm and for a length of 1 cm. This is consistent with the estimate in Section 1.2.3 that the transverse dimension may be limited by opacity to $\sim 25 \mu$m. For this $\sim 100 \mu$m annular thickness, the volume of lasing is 2.4×10^{-4} cm^3. If even 1/6 of the 13 GW (1000 J, 75 ns) input power were to be concentrated somehow into this shell, the power density deposited would be $\sim 10^{13}$ W/cm^3. This is roughly 100 times that which must be stored in the upper-laser state, according to Table 4 of Section 2.5, Chapter 2. Hence the conversion efficiency to laser-state population in this device ($\sim 1\%$) is reasonable, for a thin annulus. However, the pumping efficiency for C^{5+} at least, when defined as the measured gain product (*GL*) divided by the power density in the driver is approximately 100 times that from other experiments[31]. This is indicated in footnote "c" in Table 1 and plotted in Fig. 1 of Chapter 7.

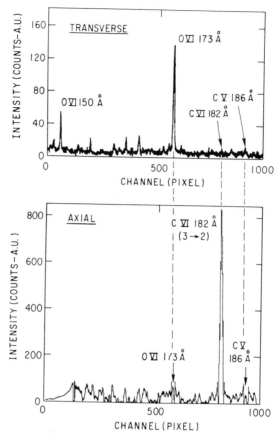

Fig. 5. Transverse and axial spectra from a carbon-disk target. (From Ref. 16.)

The long gain pulse (10–30 ns) from this novel device renders it particularly amenable to cavity operation. Unfortunately the mechanisms and detailed hydrodynamics are not yet as well understood as for the freely expanding approach. Until that improves with careful diagnostics and further modeling, the feasibility of extrapolation to other plasma conditions appropriate for shorter wavelengths will likely remain a matter of conjecture.

3. RECOMBINATION ANALYSIS EXTENDED TO PHOTOIONIZED PLASMAS

If the initial recombining ion can be produced by photoionization of a relatively cold plasma, recombination can proceed at a high rate without forced cooling. This is particularly true for ions in the rare-gas sequences, e.g., for He-, Ne-, Ar-like, ions. For these sequences the ionization potential is high and ground state population densities can be large at relatively low temperatures.

For example, in analogy to the C^{5+} example in Section 1, consider $n = 3$ to $n = 2$ lasing in He-like C^{4+} ions, pumped by recombination from hydrogenic ions. A dominant population density of He-like ground state ions in the plasma can be created[32] at a temperature of $kT \approx 20$ eV, i.e., $kT/\chi \approx 0.05$ again. Sufficient photoionization then transfers this population density into hydrogenic C^{5+} ground-state ions without further significant heating. From there, recombination pumping proceeds at the low temperature desired for rapid pumping, narrow laser lines, and high gain.

Naturally, the photon source must be very intense. Not only must the energy transferred be very high, but also the power. This is because photoionization must proceed at a higher rate than that for recombination and cascade, in order to build up a significant population inversion for lasing. The radiation should also be concentrated in the region of maximum photoionization, either by design or by filtering, in order to avoid additional heating.

There is another advantage in this approach compared to direct photoexcitation pumping analyzed in Chapter 2, Section 3. In this case, the photon is absorbed on a transition into the continuum, and no precise line-matching is required. Hence there are virtually no restrictions to particular elements. Also, the energy transfer efficiency is not hampered by an inexact overlap of line positions and profiles in transient and dynamic x-ray laser plasmas, as was the possibility for matched-line photoexcitation.

3.1 The Two-Step Pumping Sequence

This pumping scheme differs from that in Section 1 in that the recombining plasmas ions $X_0^{(i+1)+}$ are created by photoionization of a valence electron, rather than by electron-collisional ionization. The ionization process is

$$X_s^{i+} + h\nu \to X_0^{(i+1)+}, \tag{16}$$

where X_s labels the initial source ion. The entire scenario is illustrated in Fig. 6. Taken as a whole, this is seen to be a self-replenishing scheme, with the reionization controlled by the photon flux.

The recombination portion of the analysis is similar to that in the preceding section. What is treated in this section is the additional photoionization pre-process. The intent is to show that the creation of a significant initial laser state population density N_0 is possible by photoionization.

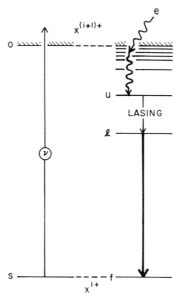

Fig. 6. Energy level diagram for pumping by electron-collisional recombination and cascade, following photoionization.

Experiments are underway for the Ne-like species, as described in Section 4. However, for a simple analysis based on hydrogenic functions which follows the above arguments, we may start with the next-simplest He-like case. We then expand to perhaps less accurate estimates for Ne-like ions. A somewhat more detailed hydrogenic-ion analysis for both Balmer-α ($n = 3–2$) as well as Lyman-α ($n = 2–1$) also has been published recently[33].

3.2. Photoionization of a He-like Plasma

For the He-like ion example, the photoionization process creating the recombining state "o" from the initial He-like ground state "s" is represented by

$$X_s^{(Z-2)+} + h\nu \rightarrow X_o^{(Z-1)+}. \tag{17}$$

This is followed by the collisional recombination process populating the u ($n = 3$) level through cascade, i.e.,

$$X_o^{(Z-1)+} + 2e \rightarrow X_n^{(Z-2)+} + e \rightarrow X_u^{(Z-2)+} \qquad (u = 3). \tag{18}$$

Lasing then takes place from $n = 3$ to $n = 2$. (For the Ne-like laser case analyzed below, $Z - 1$ and $Z - 2$ are replaced by $Z - 9$ and $Z - 10$, respectively.)

Continuing with the He-like example, we require sufficient photoionization to produce a recombining ion density N_o at least equal to the initial ion density N_s. The proportion is determined by detailed balancing [assuming collisional recombination dominates over radiative recombination (see Fig. 2)]. This condition is expressed by

$$N_s P_{pi} \geq N_o P_{cr}. \tag{19}$$

3.2.1. The Recombination Rate

He-like ions are sufficiently similar to hydrogenic that we can use a modified hydrogenic model. This is much more straightforward then deriving separate equations as was done in Section 3 of Chapter 3 for the Ne-like case. Hence, for the recombination rate P_{cr}, we use Eq. (3) of Section 1, with n' taken as 2.5 again. We substitute $(Z - 0.6)$ for the hydrogenic Z, where the 0.6 is chosen as the effective screening factor of the second electron. With this factor, the ionization potential can be written as $\chi = \mathrm{Ry}(Z - 0.6)^2$, in analogy to the hydrogenic case. We also assume again that $kT/\chi = 0.05$ and $N_e = 1 \times 10^{14} (Z - 0.6)^7$ cm^{-3}. Then the recombination rate becomes [see Eq. (6)]:

$$P_{cr} \approx 700(Z - 0.6)^8 \quad \sec^{-1}. \tag{20}$$

The conversion from Z to laser wavelength follows from the empirical expression $\lambda_{u\ell} = \lambda_{32} \approx 7000/(Z - 0.6)^2$ in Angstrom units.

3.2.2. The Photoionization Rate

For the photoionization rate P_{pi} in Eq. (19), we follow the arguments in Section 3 of Chapter 3, i.e.,

$$P_{pi} = N_v \sigma_{pi} c. \tag{21}$$

The photoionization cross-section σ_{pi} can be written as[34]

$$\sigma_{pi} = \frac{16}{3(3)^{1/2}} \frac{hc}{\mathrm{Ry}(Z - 0.6)^2} r_0 \langle g_{n\infty} \rangle n_{pi}. \tag{22}$$

The photoionization Gaunt factor $\langle g_{n\infty} \rangle$ (not to be confused with that for electron-collisional excitation in Chapter 3) for the 1s electron near the absorption edge is[34] 0.8. Then, the cross-section for the ground state ($n_{pi} = 1$) becomes

$$\sigma_{pi} = \frac{(6.3 \times 10^{-18})}{(Z - 0.6)^2} \quad \mathrm{cm}^2. \tag{23}$$

For the photon density N_v in Eq. (21), it is generally assumed that $n = 2$ to $n = 1$ lines will be used as a source. We can take the optically thick black-body approach used in Section 3 of Chapter 3 to obtain an expression [similar to Eq. (35) there] for N_v of

$$N_v = \frac{3(8\pi\rho 8\chi^3)}{(hc)^3} \frac{\Delta v}{v},$$ (24)

where $hv = 2\chi = 2 \, \mathrm{Ry}(Z - 0.6)^2$ is assumed for the driving photon energy. (We use the same Z and He-like species here as for the absorber. Actually Z will be 1–2 increments higher in order for the photon energy to exceed the ionization energy.) The initial factor of three in this expression represents a total for three possible Kα lines. The parameter ρ is the occupation index defined in Eq. (34) of Chapter 3, and again is taken to be 0.1. If we again assume $\Delta v/v = 3 \times 10^{-4}$, we arrive with the above approximations at

$$N_v = 2 \times 10^{13} \, (Z - 0.6)^6 \quad \mathrm{cm}^{-3}.$$ (25)

We can now substitute Eqs. (23) and (25) into Eq. (21) to get the photo-ionization rate

$$P_{\mathrm{pi}} = 4 \times 10^6 \, (Z - 0.6)^4 \quad \mathrm{sec}^{-1}.$$ (26)

As with photoexcitation in Chapter 3, Section 3, such an estimated photon-pumped rate may be unrealistically large compared to that obtainable experimentally. In this case, the solid angle of photon collection by the absorber is the main correction that could enter. Note again that no losses due to non-exact line matching, such as encountered in direct photoexcitation pumping, enter here.

3.2.3. Fractional Ionization

Comparing this to the recombination rate in Eq. (20) according to Eq. (19), we find that the population ratio is

$$\frac{N_o}{N_s} = \frac{(6 \times 10^3)}{(Z - 0.6)^4}.$$ (27)

In terms of the laser wavelength $\lambda_{u\ell}$ in Angstrom units, this becomes

$$\frac{N_o}{N_s} = 1 \times 10^{-4} \, \lambda_{u\ell}^2.$$ (28)

For the $Z = 6$ example used previously, $\lambda_{u\ell} = 240 \, \text{Å}$ and this ratio is ~ 6, i.e., considerably greater than the specified value of unity. For $Z > 9$ ($\lambda_{u\ell} <$

100 Å) the ratio becomes less than unity, at least according to this simple analysis. Lower densities for longer lengths and reduced recombination rates could extend this limitation to shorter wavelengths.

3.3. Photoionization of a Ne-like Plasma

For the neon-like case, we use some of the scaling relations developed in Section 1 of Chapter 3, based on a $Z - 9$ core ion.

3.3.1. The Recombination Rate

We begin by assuming a recombination temperature of $kT = 0.05\chi = 0.86(Z - 9)^{1.5}$, using Eq. (5) of Chapter 3. Also, the optimum electron density is obtained for this temperature in analogy to Eq. (13) of Chapter 3, i.e.,

$$(N_e)_{opt} = 2 \times 10^{15} (Z - 9)^{3.75} \quad cm^{-3}. \tag{29}$$

This equals 3×10^{20} cm^{-3} for the Se^{24+} ($Z - 9 = 25$) example, which is a typical value in Ne-like lasing experiments (Chapter 3, Section 2).

With these parameters, the recombination rate becomes, using Eq. (3) above:

$$P_{cr} = 2 \times 10^5 (Z - 9)^{1.5} \quad sec^{-1}. \tag{30}$$

The conversion to laser wavelength follows from $\lambda_{u\ell} = 4600/(Z - 9)$ [Eq. (11) of Chapter 3], resulting in

$$P_{cr} = \frac{6 \times 10^{10}}{\lambda_{u\ell}^{3/2}} \quad sec^{-1}, \tag{31}$$

for $\lambda_{u\ell}$ in Angstrom units. For $Z - 9 = 25$ and $\lambda_{u\ell} = 207$ Å, this becomes $P_{cr} = 2 \times 10^7$ sec^{-1}, a reasonable value.

3.3.2. The Photoionization Rate

The Ne-like photoionization cross-section for a 2p electron ($n = 2$) and a corresponding Gaunt factor of 0.9 is found from Eq. (22), substituting $\chi = 17(Z - 9)^{1.5}$ eV in place of $(Z - 0.6)^2$ Ry. The result is

$$\sigma_{pi} = \frac{1 \times 10^{-17}}{(Z - 9)^{1.5}} \quad cm^2. \tag{32}$$

Similarly, the photon density from three possible optically thick pumping lines is found from Eq. (24) for $\rho = 0.1$ to be:

$$N_v = 4.6 \times 10^{13} (Z - 9)^{4.5} \quad cm^{-3}. \tag{33}$$

The photoionization rate from Eq. (21) is then

$$P_{pi} = 1.4 \times 10^7 (Z - 9)^3 \quad sec^{-1}. \tag{34}$$

3.3.3. Fractional Ionization

With the above estimates, the specified Ne-like laser population ratio becomes, using Eq. (19),

$$\frac{N_o}{N_s} = 78(Z - 9)^{3/2}. \tag{35}$$

This appears to be much greater than unity for all Zs, at least in this simple analysis. This is most encouraging for the ongoing experiments with photoionized Ne-like ions followed by recombination for 3p-3s lasing (described in the next section).

4. RECOMBINATION EXPERIMENTS: PHOTOIONIZED PLASMAS

Creating a plasma by photoionization and subsequently forming a population inversion by recombination is more complex than is recombination in a single-species plasma, as discussed in Sections 1 and 2. It involves a second plasma and the associated interactions. An advantage is the potential for maintaining a lower temperature for increased gain on Doppler broadened lines, as explained in Section 2 of Chapter 2. There is at least one serious laboratory effort underway to achieve amplification through this combination.

4.1. Pumping Ne-like Ions by Photoionization/Recombination

4.1.1. The Processes Involved

Photoionization followed by recombination pumping is currently under study in an experiment using a z-pinch as a radiation source to ionize Ne-like Cu and Ni ions. Recombination pumping of 3p-3s transitions from F-like ions follows, for lasing in the 200–300 Å range.

Such lasing already has been very successfully demonstrated when the pumping occurs by electron-collisional excitation from the Ne-like ground state. This is discussed in detail in Section 2 of Chapter 3. The potential advantages with photoionization initiation are a much lower temperature, and possibly a larger size for increased output. For example, the electron temperature in the lasant plasma may be as low as $kT_e = 50$ eV to maintain a Ne-like ground state population. This can be compared to $kT_e \approx 500$ eV required for collisional pumping in the Ne-like copper experiments described in Section 2 of Chapter 3. Also, collisional recombination is now well established as a pumping mechanism. This is discussed in Sections 1 and 2 of this chapter.

Hence, this experiment draws on two very successful and proven features in x-ray laser research, namely 3p-3s inversions associated with 3p metastability in Ne-like ions, and rapid collisional recombination pumping.

4.1.2. The z-Pinch Photoionization Driver

To obtain sufficient radiative power, a pulsed power source such as is described in Section 4.2.2b of Chapter 2 is used. At present this is the Proto II machine at Sandia National Laboratory (SNL). In recent experiments, this device is operated at a potential of 1.2 MV and delivers 6 MA of current and 10 TW of power over a 45 ns pulse interval into a 0.125 Ohm puffed-gas z-pinch load[35-38]. Up to 15 kJ of total K-shell radiation is measured in a neon test plasma throughout a 15 ns pulse for a power density of > 1 TW/cm^2 at the center line of the collapsing cylindrical plasma. This is a very impressive 10% efficiency of conversion. It promises an extremely bright photoionization source of limited spectral breadth. This is very desirable for photoionization without undo radiative heating of the Ne-like plasma.

4.1.3. The Target Design

The target design for the SNL experiment is somewhat complex and is illustrated in Fig. 7. The laser medium (Cu or Ni in this case) is deposited on the inside of a cylindrical stagnation shell. The latter is also referred to as a "soda straw" and is constructed of low-density foam material. The outer surface is coated with the conversion material that generates the pumping x-rays. In this case it is aluminum. The purpose of the stagnation layer is to stagnate the imploding plasma, converting plasma kinetic energy into radiation. It also serves to isolate the laser rod from the imploding plasma to delay shock wave disruption of the laser medium until lasing is complete.

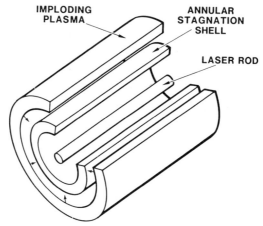

Fig. 7. Cylindrical "pie diagram" of a plasma-implosion-driven soft x-ray laser configuration. (From Ref. 38.)

Hence, the aluminum is heated by the discharge, expands, and also implodes on the foam and copper or nickel inside layer, thereby thermally exploding and heating the innermost lasant material. X-rays from the aluminum photoionize the lasant to the F-like state. Subsequently, annular x-ray lasing is expected to occur in Ne-like ions following recombination. This happens in the interval before total collapse and the onset of magnetic Rayleigh-Taylor instabilities, which could lead to an inhomogeneous lasant plasma. No lasing has been detected as of late 1989.

4.2. Photoionization/Recombination of Na^{8+} Using C^{4+} Emission

There is already other evidence on a smaller scale that photoionization can indeed lead to enhanced recombination and enhanced population of excited states. In an experiment at the Naval Research Laboratory[39], 300 eV photons (40.3 Å wavelength) from a $n = 2$ to 1 resonance transition in He-like C^{4+} ions were found to be very effective in photoionizing Li-like Na^{8+} ions with an ionization potential of 299 eV. The plasmas were created with a focused laser and a sodium-impregnated carbon target. The observed carbon intensity decreased during a period in which emission on transitions such as 3d-2p from excited sodium states increased. This was associated with enhanced populations following recombination pumping and cascading with the additional ions created by photoionization.

5. ANALYSIS OF DIELECTRONIC RECOMBINATION PUMPING

5.1. The Process

The process of dielectronic recombination differs from either radiative- or collisional-recombination (both of which involve the capture of a free electron by a stable ion), in that it includes the excitation of another initially bound electron (hence the name "dielectronic"). In this process, described by the equation

$$X_s^{(i+1)+} + e \leftrightarrow (X_0^{i+})^{**} \rightarrow (X_u^{i+})^* + hv \tag{36}$$

and illustrated in Fig. 8, a free electron attaches itself to an ion X_s in a radiationless transition. At the same time, another bound electron is excited from the ground state. This results in a temporary doubly excited quasi-bound state, i.e., a metastable complex. This is indicated by double asterisks (**) both in Eq. (36) and in Fig. 8. (Single asterisks were omitted in previous such equations for clarity.)

The doubly excited intermediate ion may decay through various channels. The most probable is the inverse of the first step [refer to Eq. (36)]. This process is autoionization, with the radiationless re-release of the captured electron. Next most likely, and of the greatest interest here, is a stabilizing

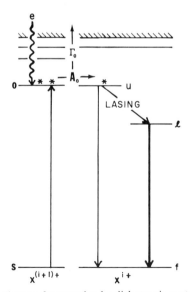

Fig. 8.   Energy level diagram for pumping by dielectronic capture and stabilization.

transition in which radiative decay of one (and ultimately both electrons) takes place to a state of sufficiently low energy to be stable. This is then a rearrangement collision, shown as the second process in Eq. (36), and results in (dielectronic) recombination into an excited state labeled "u". This may be a laser state or may be followed by further cascading into a laser state. (There is also a more remote possibility that relaxation of the doubly excited intermediate-ion state will result in release of both the electron and a photon (of variable energy) in a "radiative-Auger" rearrangement collision, with bremsstrahlung-type emission[40].)

Thus, the overall process is two-step and non-self-replenishing. The free electron must have a definite kinetic energy of value less than threshold for direct excitation of the bound electron by the amount of its final binding energy. Hence, the capture process is actually a subthreshold resonance in the overall scattering event.

There are therefore branching ratios for the intermediate state (labeled "o"). These are determined (at least in the collision-free limit) mainly by the autoionization rate (Γ_o) and by the radiative decay rate (A_o). The stabilizing radiative branching fraction $A_o/(A_o + \Gamma_o)$ for the first electron determines the degree of recombination into specific excited states of the second electron. This is the matter of primary interest here, for potential population inversion. This ratio will be very small except for $A_o \geq \Gamma_o$, and this occurs only at high Z (≥ 40).

The free-electron capture process somewhat resembles collisional-recombination pumping. However, excitation of a bound electron is also reminiscent of electron-collisional excitation pumping. Such similarities will become more apparent as we proceed in the analysis.

5.2. Analysis

5.2.1. The Recombination Fraction in He-like Ions

5.2.1a. THE DIELECTRONIC RECOMBINATION RATE. Proceeding with the analysis for lower-Z elements, the rate of population per ion of a particular upper-laser level by the "promoted" electron is given by

$$P_{dr} = N_e \langle \sigma_{dr} v \rangle \frac{A_o}{(A_o + \Gamma_o)}, \tag{37}$$

where the parameters A_o and Γ_o represent the radiative and autoionization decay rates for the intermediate state. Here $N_e \langle \sigma_{de} v \rangle$ is the dielectronic capture rate averaged over a Maxwellian electron velocity distribution [for the first reaction in Eq. (36)].

The detailed analysis below will be limited to the simplest case of $n = 3$ to $n = 2$ lasing transitions in low-Z He-like ions. This follows free-electron capture by a hydrogenic ion in the initial "source" state (s) and, subsequently, the radiative stabilization of the 2p3l intermediate state (o) by a 2p-1s transition.

The rate coefficient in Eq. (37) is obtained by the principle of detailed balancing from the inverse autoionization rate as[41,42]

$$\langle \sigma_{dr} v \rangle = \frac{h^3 \Gamma_o}{2(2\pi m k T_e)^{3/2}} \frac{g_o}{g_s} \exp\left[\frac{-\Delta E_{so}}{kT_e}\right]. \tag{38}$$

Here $g_s = 2$ and $g_o = 3$ are the statistical weights of the initial (hydrogenic-ground) state "s" and an intermediate doubly excited (**) 2p3l, $J = 1$ He-like bound state "o", respectively. The parameter ΔE_{so} is the absolute energy difference between these states. It is given approximately by the energy of the $n = 2$ to $n = 1$ Lyman-α transition of the hydrogenic ion (ignoring the $n = 3$ electron presence), i.e.,

$$\Delta E_{so} \approx \frac{3Z^2 \, \text{Ry}}{4}. \tag{39}$$

It is assumed for low to moderate Z that $\Gamma_o \gg A_o$. In this approximation, the combining of Eqs. (37) and (38) results in a cancellation of Γ_o, so that the pumping rate becomes

$$P_{dr} = N_e \frac{h^3 A_o}{2(2\pi m k T_e)^{3/2}} \frac{g_o}{g_s} \exp\left[\frac{-\Delta E_{so}}{kT_e}\right]. \tag{40}$$

Numerically, this reduces to

$$P_{dr} = \frac{(2 \times 10^{-22})}{kT_e^{3/2}} N_e A_o \frac{g_o}{g_s} \exp\left(\frac{-\Delta E_{so}}{kT_e}\right) \quad \text{sec}^{-1}, \tag{41}$$

for kT_e in eV. This relation has a strong (inverse) temperature dependence. As such, it tends to favor collisional recombination more so than electron-collisional excitation pumping.

5.2.1b. THE OPTIMUM ELECTRON TEMPERATURE.   In order to obtain a value for the optimum temperature from Eq. (40), we note that the pumping rate increases exponentially with T_e at low values and decreases as $T_e^{-3/2}$ at high values. Hence, using Eq. (39), we arrive at an optimum temperature of

$$(kT_e)_{opt} = \left(\frac{2}{3}\right)\Delta E_{so} \approx \frac{Z^2\,\mathrm{Ry}}{2} \approx 7Z^2 \quad \mathrm{eV}. \tag{42}$$

Then the ratio $\Delta E_{so}/kT_e = 3/2$, for use in Eq. (40).

5.2.1c. THE OPTIMUM ELECTRON DENSITY.   For the electron density in Eq. (40), it is sufficient for He-like ions to adopt the somewhat conservative value used in Section 1.2.1, namely $N_e = 1 \times 10^{14}\,Z^7$ cm^{-3}. When this is modified for He-like ions it becomes, approximately,

$$N_e \approx 1 \times 10^{14}\,(Z - 0.6)^7 \quad \mathrm{cm}^{-3}. \tag{43}$$

5.2.1d. THE NET PUMPING RATE.   Putting   these   values   for kT $[\approx 7(Z - 0.6)^2]$, $\Delta E_{so}/kT \approx 3/2$ and N_e into Eq. (41), along with $g_o/g_s = 3/2$ and $A_o \approx 4.7 \times 10^8\,Z^4 \approx 7 \times 10^8\,(Z - 0.6)^4$ sec^{-1} (modified hydrogenic $n = 2$ to $n = 1$), we arrive at

$$P_{dr} = 0.2(Z - 0.6)^8 \quad \mathrm{sec}^{-1}. \tag{44}$$

The Z-scaling to the 8th power here is the same as for collisional recombination [Eq. (6)]; however, the magnitude is much smaller here.

This pumping rate can be converted to laser wavelength $\lambda_{u\ell}$ for He-like $n = 3$ to $n = 2$ transitions using the same empirical expression as in Section 3.2.1, namely, $\lambda_{u\ell} = \lambda_{32} \approx 7000/(Z - 0.6)^2$ Å. The result is

$$P_{dr} = \frac{5 \times 10^{14}}{\lambda_{u\ell}^4} \quad \mathrm{sec}^{-1}. \tag{45}$$

For $Z = 6$, $\lambda_{u\ell} = 240$ Å, this becomes $P_{de} = 2 \times 10^5$ sec^{-1}. This is low in comparison with other rates derived so far.

Therefore, this process is typically considered to be a supplement to other forms of pumping at high densities. This has recently been borne out in analyses of gain experiments on $\Delta n = 0$ transitions in $Z = 34$ Ne-like ions at the Lawrence Livermore National Laboratory[43].

5.2.2. The Upper-State Density and Gain

5.2.2a. THE DEGREE OF PUMPING.   Proceeding as in Section 1.2.2, the ratio of upper (u) state density to the initial state (s) density is, for dielectronic-recombination pumping

$$\frac{N_u}{N_s} = \frac{P_{de}}{2A_{uf}}. \tag{46}$$

Here A_{uf} refers to the decay of the stabilized 1s3l upper (u) laser level in the He-like ion in this example to the final (f) $1s^2$ ground state. This can be estimated[7] by $A_{uf} = A_{31} \approx 3 \times 10^8 (Z - 0.6)^4$ sec^{-1}. Using Eq. (44), we have

$$\frac{N_u}{N_s} = 3 \times 10^{-10} (Z - 0.6)^4. \qquad (47)$$

Converting to wavelength, this becomes

$$\frac{N_u}{N_s} = \frac{2 \times 10^{-2}}{\lambda_{u\ell}^2} = 3 \times 10^{-7}, \qquad (48)$$

for $\lambda_{u\ell} = 240$ Å.

5.2.2b. INITIAL- AND UPPER-STATE DENSITIES.   For $N_s \approx N_e/3(Z - 0.6)$, we have

$$N_s = 3 \times 10^{13} (Z - 0.6)^6 \quad cm^{-3}. \qquad (49)$$

With Eq. (47), this gives

$$N_u = 9 \times 10^3 (Z - 0.6)^{10} \quad cm^{-3}, \qquad (50)$$

or

$$N_u = \frac{1 \times 10^{23}}{\lambda_{u\ell}^5} = 1 \times 10^{11} \quad cm^{-3} \qquad (51)$$

for $Z = 6$ and $\lambda_{u\ell} = 240$ Å again. This also is low in comparison with other pumping methods analyzed so far.

5.2.2c. THE GAIN COEFFICIENT.   Using Eq. (12″) in Table 1 of Chapter 2, we derive

$$G_{dr} = \frac{2 \times 10^5}{\lambda_{u\ell}^4} \quad cm^{-1}, \qquad (52)$$

for[7] $f_{\ell u} = f_{23} = 0.7$, $g_{\ell} = g_2 = 1$, $g_u = g_3 = 3$ and $F = 1/3$ again. This coefficient scales with wavelength the same as for collisional recombination. However, the magnitude is lower by approximately 10^4 times. This reflects the lower pumping rate associated in part with the more restricted energy range of the free electrons involved in the reaction. For $\lambda_{u\ell} = 240$ Å, this results in $G_{de} = 6 \times 10^{-5}$ cm^{-1}, which is definitely uninteresting in itself.

There are no known experimental results to report for direct dielectronic-recombination pumping. Hence there is no section following this on that subject.

6.   ANALYSIS OF CHARGE TRANSFER PUMPING

6.1. Atom-Ion Charge Transfer

In this method of pumping[44,45], the ion acquires an electron in an upper laser state from a neutral atom, rather than a free electron. Still, in many respects it resembles electron-collisional recombination pumping described in Section 1. Ion-ion charge transfer has a much lower cross-section than does the atom-ion process, and will only be discussed briefly at the end of this section.

As was the case for photoexcitation pumping (Section 3 and 4 of Chapter 3), there are two components involved here, the lasant and the pump. In this case however, they are mixed, and it is more difficult to control the pump medium (neutral atoms).

The dominant single-electron exchange process can be described by

$$Y + X_o^{(i+1)+} \rightarrow Y^+ + X_u^{i+}, \tag{53}$$

for a neutral perturbing atom Y impacting on a $(i + 1)$-times ionized target ion X. The process is energy-resonant between the binding energies of the electron in the ground state (g) of the atom and that of the upper-laser state of the target ion. Also, the reaction must be exothermic, i.e., the latter must exceed the former. This is diagrammed in Fig. 9.

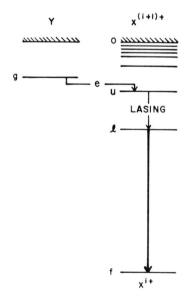

Fig. 9.   Energy level diagram for pumping by charge transfer.

In the scaling analysis below, we again use for convenience a hydrogenic model, i.e., the initial ion $X_0^{(i+1)+}$ is completely stripped and the final ion X_u^{i+} has one electron in an excited state n_u. Then the basic exothermic condition becomes

$$\frac{Z_x^2 \, \text{Ry}}{n_u^2} > \chi_y. \tag{54}$$

Here χ_y is the ionization energy of the neutral atom Y, and $\zeta = Z_x$, the nuclear charge of the X-atom. If we take $\chi_y = 1$ Ry for an estimate, the exothermic limit on n_u is

$$n_u < Z_x. \tag{55}$$

This represents the upper limit for the quantum state in the hydrogenic ion of nuclear charge Z_x for which such an atom can donate a ground-state electron exothermically.

Somewhat more precisely, Landau-Zener theory shows that the peak cross-section for the resonant charge transfer reaction occurs not at this exothermic limit, but rather when there is an energy decrement $\Delta E \approx 10\text{--}20$ eV or ~ 1.1 Ry. Then, Eq. (54) becomes

$$\frac{Z_x^2 \, \text{Ry}}{n_u^2} > \chi_y + 1.1 \, \text{Ry} \approx 2.1 \, \text{Ry}. \tag{56}$$

This leads to the "optimum" condition

$$n_u = \frac{Z_x}{1.5}. \tag{57}$$

Hence, the range is

$$\frac{Z_x}{1.5} < n_u < Z_x. \tag{58}$$

For the carbon example again, $Z_x = 6$ and the optimum final upper laser state will be $n_u = 4$, with a maximum from Eq. (55) of $n_u = 6$. This is what is observed in experiments to be described in Section 7.

6.2. Analysis

In the following analysis, we will assume that $n_u = 3$ and $n_\ell = 2$ for the laser transition, for comparison with other pumping schemes. This could evolve from the above criteria by cascade from $n = 4$–6, either spontaneously or through stimulated emission at longer wavelengths.

6.2.1. The Pumping Process [8]

6.2.1a. THE CHARGE TRANSFER CROSS SECTION.   In deriving a pumping rate P_{ct} for the charge transfer (ct) reaction, we start with the cross-section σ_{ct}. This can be approximated by[44]

$$\sigma_{ct} \approx \pi a_0^2 Z_x^2 = 9 \times 10^{-17} Z_x^2 \quad \text{cm}^2. \tag{59}$$

6.2.1b. THE NEUTRAL ATOM DENSITY.   We assume that the neutral atom density N_y is related to the ion density by $N_y = N_x/100 = N_e/300Z_x$. For the electron density we take the value used in Sections 1.2.1 and 5.2.1c, namely, $N_e = 1 \times 10^{14} Z_x^7 \text{ cm}^{-3}$. Then we have

$$N_y = 3 \times 10^{11} Z_x^6 \quad \text{cm}^{-3}. \tag{60}$$

6.2.1c. RELATIVE VELOCITY.   We can also take the relative ion/atom velocity as that of the thermal ions. We scale according to $(T_x/M_x)^{1/2}$ from $v_x = 10^7$ cm/sec for C^{6+} ions at an ion temperature of $kT_x = 200$ eV, and find

$$v_x = 10^7 \left(\frac{T_x}{200} \frac{6}{Z_x} \right)^{1/2} = 5 \times 10^6 (Z_x)^{1/2} \tag{61}$$

for $T_x = Z_x^2 \text{ Ry}/2$ in eV.

6.2.1d. THE PUMPING RATE.   The pumping rate

$$P_{ct} = N_y \sigma_{ct} v_x \tag{62}$$

now becomes, using Eqs. (59)–(61),

$$P_{ct} = 140 Z_x^{8.5} \quad \text{sec}^{-1}. \tag{63}$$

For $\lambda_{u\ell} = \lambda_{32} = 6562/Z^2$, this scales with wavelength as

$$P_{ct} = \frac{(2 \times 10^{18})}{\lambda_{u\ell}^{4.25}} = 6 \times 10^8 \quad \text{sec}^{-1}, \tag{64}$$

for $\lambda_{u\ell} = 182$ Å. Hence this pumping rate is similar to that for collisional recombination [Eq. (7)].

6.2.2. The Upper-State Density and Gain

6.2.2a. THE DEGREE OF PUMPING. The fractional density pumped is found from

$$\frac{N_u}{N_o} = \frac{P_{ct}}{2A_{uo}} = \frac{500}{\lambda_{u\ell}^{2.25}} = 0.004, \tag{65}$$

for $\lambda_{u\ell} = \lambda_{32} = 182$ Å. Also here we use[7] $A_{uo} = A_{31} = 5.6 \times 10^7 \, Z^4 \, \sec^{-1}$ once again.

6.2.2b. INITIAL- AND UPPER-STATE DENSITIES. Taking $N_o = N_e/3Z_x$ ($N_o = N_x$ here), the initial-ion density becomes

$$N_o = 3 \times 10^{13} \, Z_x^6 \quad \text{cm}^{-3}. \tag{66}$$

Combining this equation with Eq. (65) and converting to wavelength, the upper-laser-state density becomes

$$N_u = \frac{(4 \times 10^{27})}{\lambda_{u\ell}^{5.25}} = 6 \times 10^{15} \quad \text{cm}^{-3}, \tag{67}$$

for $\lambda_{u\ell} = 182$ Å.

6.2.2c. THE GAIN COEFFICIENT. The gain coefficient is now found from Eq. (12″) in Table 1 of Chapter 2 to be for $f_{\ell u} = 0.64$,

$$G_{ct} = \frac{(2 \times 10^{10})}{\lambda_{u\ell}^{4.25}} = 4 \quad \text{cm}^{-1}, \tag{68}$$

at $\lambda_{u\ell} = 182$ Å. This is close in both magnitude and scaling to that for recombination pumping. However, in this case there is an additional independent variable in the neutral atom density. If it can be controlled and perhaps injected at a density greater than the value of $N_x/100$ assumed above, the gain will scale to even larger values.

6.3. Ion–Ion Charge Transfer

A major practical problem with the atom/ion charge transfer pumping approach is to maintain the atoms in a neutral state congruent with or adjacent to a high-temperature lasant plasma.

This problem would be significantly lessened if one could use ion-ion interactions, i.e., the donor atom Y in Eq. (53) becomes Y^{m+}. Such reactions do take place, but at a much smaller cross-section and for a much larger impact velocity in the presence of strong Coulomb repulsion. The available data are sparse for such reactions with multiple ionization. However, we can gain some insight into the situation from extrapolating calculations[46] for hydrogenic "donor" ions ($Y^{(Z_y - 1)+}$) reacting with accelerated alpha particles (He^{2+}) or heavier fully ionized receptive (lasant) ions (X^{Z_x+}) of charge Z_x. It is shown in Ref. 46 that the peak cross-section scales approximately as

$$\sigma_{iict}(i - i) \approx 3.5 \times 10^{-17} \frac{Z_x^2}{Z_y^6} \quad cm^2, \tag{69}$$

This becomes

$$\sigma_{iict} \approx 5.5 \times 10^{-19} Z_x^2 \quad cm^2, \tag{70}$$

even for the lowest-Z He^+ donors. This is almost 200 times lower than for neutral donors, as given in Eq. (59).

According to Ref. 46, this peak cross-section occurs for a normalized velocity of

$$v_x(i - i) = 2.2 \times 10^8 \, Z_y \quad cm/sec. \tag{71}$$

Such velocities would be very high for bulk lasant plasmas at the densities required for gain. The velocity in Eq. (71) translates into a peak energy of

$$E_{pk}(i - i) \approx 50 Z_x Z_y^2 \quad keV. \tag{72}$$

This also reaches an unrealistic level of 1.2 MeV for the above parameters. Hence ion-ion charge transfer pumping does not appear to be a viable possibility for pumping plasma x-ray lasers, even in moderate-Z ions.

7. EXPERIMENTS WITH CHARGE TRANSFER PUMPING

There have been a number of proposals for charge-transfer-pumped x-ray lasers. However, as yet there appears to be only one successful gain experiment. As with resonant photoexcitation pumping experiments described in Section 4 of Chapter 3, there is the added complexity here of two media that must properly interact for pumping. Also, one of these must not adversely affect the necessary ionization balance in the other. In the present case the problem is to maintain a sufficient density of cold, neutral donor-atoms in the immediate vicinity of receptor-ions in the hot, ionized lasant plasma.

7.1. Plasma/Barrier Interaction Experiments

The most popular concept has been to create a rapidly streaming receptor plasma which then collides with a dense neutral donor medium such as a solid wall[47,48] or a high pressure gas[49].

7.1.1. Lyman-α Lasing Following Proton/Sodium-Atom Interactions

Gain on the Lyman-α line of hydrogen at 1216 Å has been measured for proton acceptors from a plasma gun and neutral sodium donor atoms in a vapor cloud[50]. A single 80%-reflecting mirror is used to ascertain the presence of net amplification. A gain coefficient of $G = 1.4$ cm^{-1} was reported. It appears from available literature that the further lasing experiments planned were not carried out.

7.1.2. Fluorescence from Plasma/Gas Interactions

Fluorescence has been measured also in hydrogenic and He-like ions of lithium, boron, carbon, and nitrogen[49]. The most complete results are for C^{5+} ions, which have captured an electron from neutral carbon atoms. In that case, the initial stripped C^{6+} plasma is formed by a focused laser beam onto a graphite slab target. Expansion into a background gas is used to attentuate the ion flow velocity to an optimum value for the charge transfer process. Preferential population into $n = 4$ and $n = 5$ levels in agreement with theory is measured by resonance-line fluorescence.

A density of $\sim 10^{13}$ cm^{-3} consistent with theoretical requirements is deduced from the data on the rapidly flowing neutral carbon atoms[51] found in the region of interaction. These fluorescence experiments were performed at low density, where the quantitative spectroscopy required would not be hampered by opacity effects. For that reason, translating the results to increased density closer to the target gives less definitive results.

7.2. High-Energy Ar$^+$ Beam Impacting a Carbon Target

There is a novel proposal at least for an experiment that involves a traveling argon-ion beam and the pumping of an innershell inversion by the ionizing impact with neutral carbon atoms[52]. It is suggested that $n = 2$ L-shell vacancies can be created in argon with a high preferential probability through resonance interaction with the K-shell of the carbon atoms, a process that resembles charge transfer[53]. This would then lead to a population inversion and gain at a photon energy of 224 eV ($\lambda_{u\ell} = 55$ Å). It is proposed that the collisions take place in a very thin (~ 500 Å-thick) carbon film. As an example, a beam of 30 keV energy and 0.5 Å current concentrated onto an area of 0.24 cm^2 of the foil is predicted to yield an upper-laser-state density of $N_u = 2.6 \times 10^{13}$ cm^{-3} and finally a gain coefficient of $G = 0.054$ cm^{-1}.

The suggested design involves a pulsed pumping beam in a traveling-wave mode. This beam then propagates along the gain medium in synchronization with the amplified wave packet. An overall length of about 1 m is required for ASE operation. Pulsed ion beams with current densities as high as 1 kA/cm^2 appear to be possible[54]. This conceivably could enhance the gain and ultimate output of such a scheme over that originally estimated.

REFERENCES

1. L. I. Gudzenko, and L. A. Shelepin, *Sov. Phys. Doklady* **10**, 147 (1965).
2. G. J. Pert, *J. Phys. B* **9**, 3301 (1976) and **12**, 2067 (1979).
3. S. Suckewer and H. Fishman, *J. Appl. Phys.* **51**, 1922 (1980).
4. V. A. Boiko, F. V. Bunkin, V. I. Derzhiev and S. I. Yakovlenko, *IEEE J. Quant. Electron.* **QE-20**, 206 (1984).
5. H. R. Griem, "Plasma Spectroscopy" (McGraw-Hill Publ. Co., New York, 1964).
6. R. C. Elton, "Atomic Processes," Chapt. 4 in "Methods of Experimental Physics," Plasma Physics, Vol. 9, H. R. Griem and R. H. Lovberg, eds. (Academic Press, New York, 1970).
7. W. L. Wiese, M. W. Smith and B. M. Glennon, "Atomic Transition Probabilities: Vol. 1, Hydrogen through Neon," Report No. NSRDS-NBS 4 (U.S. Government Printing Office, Washington, DC, 1966).
8. R. C. Elton, *Opt. Engr.* **21**, 307 (1982).
9. R. C. Elton, "Parameter regimes for x-ray lasing in plasmas," *Comments At. Mol. Phys.* **13**, 59–67 (1983). Because of printing error, there is some confusion between "ell" and "one" in this paper. The author will provide a marked copy upon request.
10. R. C. Elton, J. F. Seely and R. H. Dixon, AIP Conf. Proc. No. 90, T. J. McIlrath and R. R. Freeman, eds., p. 277 (Am. Inst. of Physics, New York, 1982).
11. G. J. Pert and S. A. Ramsden, *Optics Comm.* **11**, 270 (1974).
12. F. E. Irons and N. J. Peacock, *J. Phys. B* **7**, 1109 (1974).
13. R. J. Dewhurst, D. Jacoby, G. J. Pert and S. A. Ramsden, *Phys. Rev. Letters* **37**, 1265 (1976).
14. M. H. Key, C. L. S. Lewis and M. J. Lamb, *Optics Comm.* **28**, 331 (1979).
15. D. Jacoby, G. J. Pert, S. A. Ramsden, L. D. Shorrock and G. J. Tallents, *Optics Comm.* **37**, 193 (1981) and *J. Phys. B* **15**, 3557 (1982).
16. S. Suckewer, et al., *Phys. Rev. Letters* **55**, 1753 (1985), and **57**, 1004 (1986).
17. J. F. Seely, C. M. Brown, U. Feldman, M. Richardson, B. Yaakob and W. E. Behring, *Optics Comm.* **54**, 289 (1985).
18. C. Chenais-Popovics, et al., *Phys. Rev. Letters* **59**, 2161 (1987).
19. O. Willi, et al., in "Atomic Processes in Plasmas," AIP Conf. Proc. No. 168, A. Hauer and A. L. Merts, eds., p. 115 (American Institute of Physics, New York, 1988).
20. C. L. S. Lewis, *Plasma Phys. and Controlled Fusion* **30**, 35 (1988).

21. Y. Kato, P. R. Herman, T. Tachi, K. Shihoyama, K. Kamei and H. Shiraga, Proc. Int. Symposium Short Wavelength Lasers, C. Yamanaka, ed., p. 57 (Springer-Verlag, Tokyo, 1988); also *IEEE Trans. Plasma Science* **16**, 520 (1988).
22. D. L. Matthews, et al., *Appl. Phys. Letters* **45**, 226 (1984).
23. V. A. Bhagavatula and B. Yaakobi, *Optics Comm.* **24**, 331 (1978).
24. V. A. Boiko, et al., *Sov. J. Quant. Electron.* **14**, 1113 (1984).
25. S. Suckewer, C. Keane, H. Milchberg, C. H. Skinner and D. Voorhees, in "Laser Techniques in the Extreme Ultraviolet," AIP Conf. Proc. No. 119, S. E. Harris and T. B. Lucatorto, eds. pp. 55, 379 (American Institute of Physics, New York, 1984).
26. Ya. Kononov, K. N. Koshelev, Yu. A. Levykin, Yu. V. Sidel'nikov and S. S. Churilov, *Sov. J. Quant. Electron.* **6**, 308 (1976).
27. P. Jaegle, G. Jamelot, A. Carillon, A. Klisnick, A. Sureau and H. Guennou, *J. Opt. Soc. Am. B* **4**, 563 (1987).
28. J. C. Moreno, H. R. Griem, S. Goldsmith and J. Knauer, Phys. Rev. A **39**, 6033 (1989).
29. C. H. Skinner, et al., *IEEE Trans. on Plasma Science* **16**, 512 (1988); also D. Kim, et al., *SPIE Proceedings* **875**, 20 (1988) and *J. Opt. Soc. Am.* **6**, 115 (1989).
30. G. Jamelot, et al., *IEEE Trans. Plasma Science* **16**, 497 (1988).
31. M. H. Key, *J. de Physique* **49**, C1-135 (1988).
32. T. F. Stratton, in "Plasma Diagnostic Techniques," R. H. Huddlestone and S. L. Leonard, eds., p. 391 (Academic Press, New York, 1965).
33. D. G. Goodwin and E. E. Fill, *J. Appl. Phys.* **64**, 1005 (1988).
34. C. W. Allen, "Astrophysical Quantities," Second Edition, p. 90 (Athlone Press, London, 1963).
35. E. J. McGuire, et al., and M. K. Matzen, et al., *J. de Physique* **C6**, 81 and 135, respectively (1986).
36. M. K. Matzen, et al., Report No. SAND85-1151 (Sandia National Laboratory, Albuquerque, New Mexico 87185, July 1986).
37. M. K. Matzen, et al., *SPIE Proceedings* **688**, 61 (1986).
38. T. W. Hussey, M. K. Matzen, E. J. McGuire and H. E. Dalhed, Report No. SAND88-0764 (Sandia National Laboratory, Albuquerque, New Mexico 87185, May 1988).
39. R. C. Elton and R. H. Dixon, *Phys. Rev. A* **28**, 1886 (1983).
40. R. C. Elton and L. J. Palumbo, *Phys. Rev. A* **9**, 1873 (1974).
41. S. M. R. Ansari, G. Elwert and P. Mucklich, *Z. Naturforsch.* **25a**, 1781 (1970).
42. A. H. Gabriel and T. M. Paget, *J. Phys. B* **5**, 673 (1972).
43. B. L. Whitten, A. U. Hazi, M. H. Chen and P. L. Hagelstein, *Phys. Rev. A* **33**, 2171 (1986).
44. A. V. Vinogradov and I. I. Sobel'man, *Sov. Phys. JETP* **36**, 1115 (1973).
45. L. P. Presnyakov and A. D. Ulantsev, *Sov. J. Quant. Electron.* **4**, 1320 (1975); also L. P. Presnyakov and V. P. Shevel'ko, *JETP Letters* **13**, 203 (1971).
46. M. Lal, M. K. Srivastava and A. N. Tripathi, *Phys. Rev. A* **26**, 305 (1982).
47. M. O. Scully, W. H. Louisell and W. B. McKnight, *Opt. Commun.* **9**, 246 (1973).
48. D. A. Copeland, M. Mahr and C. L. Tang, *IEEE J. Quant. Electron.* **QE-12**, 665 (1976).
49. R. H. Dixon and R. C. Elton, *Phys. Rev. Letters* **38**, 1072 (1977); R. H. Dixon, J. F. Seely and R. C. Elton, *Phys. Rev. Letters* **40**, 122 (1978); R. C. Elton, T. N. Lee, R. H. Dixon, J. D. Hedden and J. F. Seely, in "Laser Interaction and Related Plasma Phenomena," Vol. 5, H. J. Schwarz, H. Hora, M. J. Lubin and B. Yaakobi, eds., p. 135 (Plenum Press, New York, 1981).
50. R. Tkach, H. Mahr, C. L. Tang and P. L. Hartman, *Phys. Rev. Letters* **45**, 542 (1980).
51. R. H. Dixon, R. C. Elton and J. F. Seely, *Optics Comm.* **45**, 397 (1983).
52. R. A. McCorkle, *Phys. Rev. Letters* **29**, 982 (1972).
53. F. W. Saris, *Physica* (Utrecht) **52**, 290 (1971), and other references included here to previous work.
54. R. B. Miller, "Intense Charged Particle Beams," p. 19 (Plenum Press, New York, 1982).

Chapter 5

Pumping by Ionization of Atoms and Ions

1. ANALYSIS OF ELECTRON-COLLISIONAL IONIZATION PUMPING

In this chapter we will follow the order of Chapter 3, beginning with ionization induced by plasma electrons followed by photons. Lastly, we will discuss Auger effects in x-ray laser pumping.

1.1. The Process

In this method of pumping, an innershell bound electron is released through ionization, following an inelastic collision between a free electron and an ion. This serves to create an atom in the next higher stage of ionization. Such a product ion could be in an excited state that is capable of generating a population-density inversion and gain. This is a less selective and less controllable process than photoionization, which is discussed in Section 2. Nevertheless, it has received some attention.

This process can be described by the equation

$$X_o^{i+} + e \rightarrow X_o^{(i+1)+}[L], \tag{1}$$

where the bracketed [L] represents a hole in a particular shell ($n = 2$ here) in x-ray notation. This is also illustrated in the energy level diagram of Fig. 1. A population-density inversion is created in those ions for which the outer-shell electron is in a higher excited state, as indicated ($n = 3$, M-shell) in the figure.

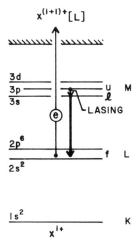

Fig. 1.   Energy level diagram for pumping by electron-collisional ionization.

1.2. An Example: Ionization of Na-like Atoms and Ions

The most familiar application of this process is with the Ne-like 3p-3s laser[1]. As described in Section 2.3.7e of Chapter 3, this process was invoked to explain the anomaly in gain coefficients for different J-sublevels in Ne-like ions. In that example, the Na-like ion in the $1s^2 2s^2 2p^6 3p$ state (highly populated at high density) is ionized into the Ne-like $1s^2 2s^2 2p^5 3p$ excited state, with 3p-3s lasing following. This scheme depends upon a high concentration of Na-like ions.

1.2.1. The Ionization Rate Coefficient

The ionization rate coefficient S_{ou} has been calculated[2] (for a Maxwellian velocity distribution) to be as large as that for direct excitation from the ground state of the Ne-like ion. We can demonstrate this quite simply here using a convenient analytical formula[3] for outershell ionization (i.e., ignoring the $n = 3$ electron in the present case), namely,

$$S_{ou} = \frac{2.5 \times 10^{-6}}{\chi^{3/2}} \frac{\mathcal{N}_e (kT/\chi)^{1/2}}{1 + (kT/\chi)} \exp\left(\frac{-\chi}{kT_e}\right) \quad cm^3 \ sec^{-1}. \tag{2}$$

Here $\mathcal{N}_e \ (=6)$ denotes the number of equivalent electrons for the 2p shell. For the ionization potential in eV, we can assume from $L_{II,III}$ x-ray absorption edge data[4] an approximate scaling for medium-Z elements of

$$\chi \approx 2.3(Z - 9)^2 \quad eV. \tag{3}$$

For $\chi/kT = 3$, as was assumed for the Ne-like example in Eq. (5) of Chapter 3, the ionization rate coefficient in Eq. (2) becomes

$$S_{ou} = \frac{9 \times 10^{-8}}{(Z - 9)^3} = 6 \times 10^{-12} \quad cm^3 \ sec^{-1}, \tag{4}$$

the latter value being for the Se^{24+} laser example ($Z - 9 = 25$). This is approximately equal to the rate coefficient for electron-collisional excitation pumping of the 3p level in selenium, given following Eq. (8) of Chapter 3.

1.2.2. Comparison with Collisional Excitation

Hence, the electron-collisional ionization process can strongly influence the Ne-like pumping if the total density of Na-like ions is as large or even larger than that for Ne-like ions. (Actually, even if these densities were equal, the $2p^6 3p$ initial state assumed above would have a density of approximately 1/3 the total Na-like ion density, according to a statistical distribution between 3s, 3p and 3d levels.) Such equivalent densities can occur in practice in a transient plasma. The necessary condition is that sequential ionization proceed slower than the heating, such that the ion species lag the steady-state value in time.

So far the importance of innershell ionization in the Ne-like laser experiments is conjecture. At present, a large Na-like ion concentration is not expected[5] to be present in the selenium experiments described in Section 2 of Chapter 3. Hence, there are no specific pumping experiments to warrant a separate section on experiments, which normally would follow here.

2. ANALYSIS OF ION-COLLISIONAL PUMPING OF IONS

Direct ion-impact collisional excitation with (or without) ionization for pumping x-ray lasers can be described by

$$Y^{m+} + X_0^{(i-1)+} \rightarrow Y^{m+} + X_u^{i+}. \tag{5}$$

This is illustrated in Fig. 2.

This process has received little attention, due in part to the greatly enhanced energy required (for similar velocities) for ions of mass at least 1800 times larger that of electrons. Also, the Coulomb repulsion significantly raises the impact energy required, much as for ion-ion charge transfer discussed in

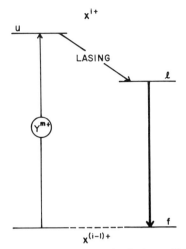

Fig. 2.   Energy level diagram for pumping by ion-collisional ionization.

Section 6.3 of Chapter 4. There are actually less data available for $\Delta n \geq 1$ ion-collisional excitation of possible laser states than there are for ion-ion charge transfer.

A reaction for which there are data available to illustrate this point is for pure excitation, i.e.,

$$p + He^+ \rightarrow He^+ \quad (n = 2). \tag{6}$$

For this reaction, a cross-section of

$$\sigma_{ce}(i - i) = 2 \times 10^{-18} \text{ cm}^2 \tag{7}$$

has been calculated[6] at an impact energy of 14 keV, or a photon velocity of 2×10^8 cm/sec. This cross-section is still on the rise, and therefore may not be a peak value. Therefore, it will probably reach values and scaling factors similar to those for ion-ion charge transfer. In any case the energy (and velocity) are again unreasonably high for bulk-plasma x-ray laser conditions.

There are no known experiments in the area of direct pumping by ion-ion collisions. Hence there is no section on such a topic.

3. ANALYSIS OF PHOTOIONIZATION PUMPING

In this section we discuss direct photoionization pumping of population inversions. Photoionization can also be a secondary process. An example of this is recombination pumping of a plasma state created initially by photo-ionization, as discussed in Sections 3 and 4 of Chapter 4.

3.1. The Processes Involved

3.1.1. Photoionization

Photons in the proper energy range impinging upon an atom or ion can lead to the preferential removal of innershell electrons. That such specific innershell ionization is indeed possible is well known from x-ray fluorescence. There, innershell holes are created leading to conventional x-ray line emission.

Such a creation of an innershell hole can lead to a population inversion with respect to an outershell or valence electron. This process is described by

$$X_o^{i+} + h\nu \rightarrow X_u^{(i+1)+}[L] + e. \tag{8}$$

In this example, $[L]$ represents an $n = 2$ or L-shell vacancy, in x-ray notation.

3.1.2. Lasing Modes

Such innershell lasing transitions are usually self-terminating in a time determined by the relaxation time for filling the lower level. This filling may take place either through spontaneous emission from the upper-laser state or through collisional excitation from the ground state, when it is close. For the neutral sodium example described below, the radiative relaxation time for the 372-Å 3s-2p transition is 0.4 ns. It becomes rapidly shorter if scaled to shorter wavelengths. For example, such an innershell-hole lifetime is in the picosecond (10^{-12} sec) to femtosecond (10^{-15} sec) range for ≈ 10-Å soft x-rays and for K-shell x-rays in the Angstrom wavelength ranges, respectively.

Hence it has not seemed reasonable to search for lasing in conventional x-ray data, even if such self-terminating population inversions were somehow created. However, an alternative possibility for creating quasi-cw inversions by rapid Auger cascading is described in Sections 3 and 5.

3.1.3. Possible Auger Losses

Whenever innershell vacancies are created, there exists the possibility of also removing outershell electrons during the processes of creating and filling this vacancy. This depletes the population inversion density and quenches the gain. Besides a direct collisional ionizing collision, this may be due to an Auger transition when there exist multiple outershell electrons. The latter can be avoided if there is only one outershell electron, as in the example of sodium (or Na-like or other alkali) species, which is analyzed below.

3.2. Lasing in Na and Na-like Ions

An early proposal based on this approach is related to Fig. 3. This suggestion[7] involves photoionization of an innershell 2p L-shell electron in sodium ($1s^2 2s^2 2p^6 3s$ configuration) by a photon of energy > 38 eV. This can lead potentially to self-terminating (in 400 ps, as mentioned above) population inversion and gain. This would occur for the 3s-2p transition at a wavelength of 372 Å (or an equivalent photon energy of 33 eV). It was shown in Ref. 7 that the photoionization cross-section for a $n = 2$ electron is almost 100 times larger than that for the 3s valence electrons. This advantage exists over a fairly broad photon energy range, beginning at the 38 eV (326 Å) threshold and extending to well over 100 eV. This particular proposal was chosen as

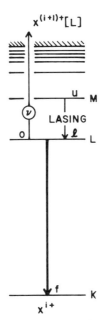

Fig. 3. Energy level diagram for pumping by photoionization.

an example for the more detailed analysis which follows. Following that, some alternate proposals will be described.

3.2.1. The Pumping Process

In this analysis we will draw on previous formulations, particularly from Section 3 of Chapter 3 on photoexcitation and Section 3 of Chapter 4 on photoionization-recombination. Comparisons will be made where possible, for further insight. We begin by defining the pumping rate

$$P_{pi} = N_v \sigma_{pi} c \quad \sec^{-1}, \tag{9}$$

for a photon density N_v and a cross section σ_{pi}.

3.2.1a. THE REACTION CROSS-SECTION.   For the photoionization cross-section, we use a modified hydrogenic expression at an energy near the absorption edge, adapted from Eq. (22) of Chapter 4, namely,

$$\sigma_{pi} = \frac{16}{3(3)^{1/2}} \frac{1}{(\xi Z_H)^2} \frac{r_0}{R_\infty} \langle g_{n\infty} \rangle n_{pi} \quad cm^2. \tag{10}$$

Here the Rydberg wavenumber constant is given by $R_\infty = Ry/hc = 109737.3$ cm^{-1}. Also, n_{pi} represents the principal quantum number for the bound electron undergoing photoionization (e.g., $n_{pi} = 2$ for the sodium example). The parameter $\langle g_{n\infty} \rangle$ is the photoionization Gaunt factor, equal to 0.95 or 0.89 for 2s or 2p electrons, respectively[8]. The correction factor $\xi > 1$ accounts for the additional Coulomb charge due to an incompletely screened nucleus. With these substitutions, the photoionization cross-section for a 2p electron can be described by

$$\sigma_{pi} = \frac{1.4 \times 10^{-17}}{(\xi Z_H)^2} \quad cm^2. \tag{11}$$

This agrees reasonably well with the $n = 2$ threshold value for sodium[7], using $\xi = 1.3$. It is, however, 200 times smaller than that for hydrogenic photoexcitation, given in Eq. (31) of Chapter 3, even though both scale hydrogenically as Z^{-2}. This reduced cross-section will be more than compensated for by an ~ 3000 times larger bandwidth for the continuum-pumped photoionization in this method, as discussed below.

It will be useful later to have this cross-section expressed in terms of the threshold ionization potential $\chi = (\xi Z)^2 R_\infty hc$ for a hydrogenic ion, i.e.,

$$\sigma_{pi} = \frac{16}{3(3)^{1/2}} \frac{hc}{\chi} r_0 \langle g_{n\infty} \rangle n_{pi}$$

$$= \frac{(1.9 \times 10^{-16})}{\chi} \quad cm^2, \tag{12}$$

for χ in eV and $n_{pi} = 2$ again.

3.2.1b. THE PUMPING PHOTON DENSITY. For the photon density N_ν, we return to the blackbody formulism developed in Section 3.1.1 of Chapter 3. From Eq. (35) of that section, we have (for $\lambda = c/\nu$):

$$N_\nu = \frac{8\pi\rho\nu^3}{c^3} \frac{\Delta\nu}{\nu}. \tag{13}$$

In the present case of photoionization we can assume $\Delta\nu/\nu \approx 1$, instead of the small value for Doppler line widths used for resonant photoexcitation pumping. Also, the frequency ν here corresponds to a mean photon energy ~ 2 times the threshold photoionization energy for the $n = 2$ electron, i.e., $h\nu \approx 2\chi$. Hence, the photon density becomes

$$N_\nu = \frac{64\pi\rho\chi^3}{(hc)^3} = 1 \times 10^{14} \, \rho\chi^3, \tag{14}$$

for χ in eV.

The ionization potential can be obtained for sodium and Na-like ions of nuclear charge Z from x-ray $L_{II,III}$-edge data[4,9], and extrapolates for low Z approximately as

$$\chi \approx 3(Z - 9)^2 \quad eV. \tag{15}$$

Here $Z - 9$ represents a partially screened 2p electron in this sequence. Inserting this into Eq. (14) and setting $\rho = 0.1$ according to the argument following Eq. (35) of Chapter 3, we arrive at

$$N_\nu = 3 \times 10^{14} (Z - 9)^6 \quad photons/cm^3. \tag{16}$$

3.2.1c. THE PUMPING RATE. Substituting Eq. (15) into Eq. (12) and inserting this along with Eq. (16) into Eq. (9), we arrive at an expression for the pumping rate

$$P_{pi} = 5 \times 10^8 (Z - 9)^4 \quad sec^{-1}. \tag{17}$$

For $\lambda_{u\ell} = 1300/(Z - 9)^{1.5}$ in Angstrom units, we can express this pumping rate in terms of the laser wavelength $\lambda_{u\ell}$ as

$$P_{\text{pi}} = \frac{(1 \times 10^{17})}{\lambda_{u\ell}^{8/3}} \quad \text{sec}^{-1} \tag{18a}$$

or

$$= 1 \times 10^{11} \quad \text{sec}^{-1}. \tag{18b}$$

The latter value is for lasing on a 3s-2p self-terminating transition in a Ne-like ($Z = 13$) Al^{3+} ion, where $\lambda_{u\ell} = 160$ Å. This is 14 times larger than was $P_{\text{pe}} = 7 \times 10^9$ sec^{-1} for photoexcitation pumping of the $n = 3$ to $n = 2$ laser transition in the C^{5+} hydrogenic ion at $\lambda_{u\ell} = 182$ Å [see Eq. (38) of Chapter 3]. Note that the scaling to the fourth power of the effective charge is the same for both.

3.2.2. The Upper-State Density and Gain

3.2.2a. THE DEGREE OF PUMPING. The fractional density pumped is again found in equilibrium to be

$$\frac{N_u}{N_o} = \frac{P_{\text{pi}}}{A_{\text{uf}}} = 1 \times (Z - 9)^{1.5} \tag{19a}$$

or

$$= \frac{1300}{\lambda_{u\ell}} = 8, \tag{19b}$$

for $\lambda_{u\ell} = 160$ Å again. Here we have assumed that $A_{\text{uf}} = A_{\text{uo}} \approx 5 \times 10^8$ $(Z - 9)^{2.5}$ sec^{-1} from compiled data[10]. The density ratio in Eq. (19) is probably optimistically large when one considers the less than perfect photon coupling. This was also the case for matched-line photoexcitation pumping in Section 3 of Chapter 3.

3.2.2b. THE INITIAL- AND UPPER-STATE DENSITIES. For the initial-state density we use $N_o = N_e/3(Z - 10) \approx N_e/2(Z - 9)$ for consistent scaling. For the electron density, we assume $N_e = 4 \times 10^{15} (Z - 9)^{3.75}$ cm^{-3} from Eq. (13) of Chapter 3. This was derived as a practical value whereby 3s-3p interactions do not dominate over 3s-2p decay (the laser transition here). For the Al^{3+} example, $Z - 9 = 4$ and $N_e = 7 \times 10^{17}$ cm^{-3}. The initial density then becomes

$$N_o = 2 \times 10^{15} (Z - 9)^{2.75} \quad \text{cm}^{-3}. \tag{20}$$

Substituting this into Eq. (19a) and converting to laser wavelength results in

$$N_u = \frac{(1 \times 10^{24})}{\lambda_{u\ell}^{2.8}} \tag{21a}$$

or

$$= 7 \times 10^{17} \quad cm^{-3}, \tag{21b}$$

for $\lambda_{u\ell} = 160$ Å again. This is large.

3.2.2c. THE GAIN COEFFICIENT.   We next derive a photoionization gain coefficient, again using Eq. (12″) from Table 1 in Chapter 2:

$$G_{pi} = \frac{(6 \times 10^{5})}{\lambda_{u\ell}^{1.8}} \quad cm^{-1}, \tag{22a}$$

or

$$= 70 \quad cm^{-1}. \tag{22b}$$

The latter value is for lasing at a wavelength of $\lambda_{u\ell} = 160$ Å in Ne-like Al^{3+} ions. Again, this is most likely an optimistic value because of losses in photon coupling. However, it is still very encouraging for photoionization pumping of ASE. It is also approximately 170 times as large as the original estimate[7] of 40 dB/m, which is equivalent to $I/I_o = 10^4 = \exp(GL)$ or $G = 0.09$ cm^{-1} for a length $L = 1$ m (see Chapter 2, Section 2.2.1a.). This difference can be understood by a 180 times lower density proposed for a heat pipe sodium vapor source in that early publication.

3.3. Other Photoionization Schemes

3.3.1. Photoionization in Lithium, with Photo-Triggering

The sodium scheme is applicable to other alkali atoms and alkali-like ions, as well as alkali-earths[11]. In particular, lithium atoms and Li-like ions have been proposed, where a 1s electron is photoionized leaving a 1s2s Li$^+$ ion[12,13] with a [K] hole. After population buildup and storage in this electric dipole-metastable state (also energetically stable against autoionization), the electrons are rapidly transferred to the 1s2p state by a tunable, long-wavelength ($\sim$9580 Å) dye laser. Population inversion and self-terminating (in $\sim$40 ps) gain are predicted for the 2p-1s transition at 200 Å. This triggering procedure is in essence a stimulated-resonant-Raman, anti-Stokes process[13].

The required traveling-wave pumping occurs somewhat naturally with an axial triggering-laser beam. An added advantage to this scheme with a separate triggering laser is the straight and narrow guiding channel promised. This does require the establishment of a population inversion between the metastable state compared and the ground state, which could be quite challenging.

A proposed variation of this approach uses a $1s2s2p$ ^4P quartet metastable storage level[14]. From this level, transfer to a $1s2p^2$ ^2P level is to be accomplished with a 2949-Å tunable laser. The strength of the spin-changing transition is a question affecting the overall viability of this idea.

3.3.2. Photoionization Pumping of Neon Atoms

A similar photoionization scheme involves the innershell ionization with 500 eV photons of 2s electrons in neon atoms, starting with the $1s^22s^22p^6$ ground configuration. Lasing would then take place on a 2p-2s transition[15] at a photon energy of 27 eV, or a wavelength of 460 Å.

3.3.3. Photoionization Pumping in the Zinc Sequence

Encouraged by successful visible lasing in Cu-like Zn^+ ions[16], extrapolation along the isoelectronic sequence to I^{24+} has been proposed[17]. Lasing at a wavelength near 25 Å is predicted. Photoionization pumping would be used to remove a 3d innershell electron from the initial Zn-like $3d^{10}4s^2$ ground configuration in I^{23+}. Lasing would then occur between the $3d^94s^2$ and $3d^{10}4p$ levels, followed by rapid depletion to the $3d^{10}4s$ ground state of the Cu-like ion. A gain of 2 cm^{-1} is predicted. Unfortunately the lasing is self-terminating in a period of approximately 10 ps, as the lower-laser level fills by collisional excitation from the final ground state. Hence, a very short-pulsed pumping source is required. This is an example of the additional quenching encountered for lasing states close to a highly populated state, at the high densities required for ASE in the x-ray region.

3.3.4. Quasi-cw X-Ray Operation with Auger Vacancy-Filling

All of the above photoionization schemes suffer from being self-terminating in very short times, thereby requiring extremely precise traveling-wave pumping. There is a proposal to circumvent this problem with photoionization

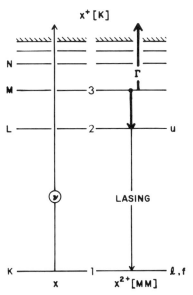

Fig. 4. Energy-level schematic for K-shell photoionization followed by an Auger cascade for cw operation through line shifting.

from a deep innershell, using an Auger cascade to shift the absorption line[18]. The essential levels are diagrammed in Fig. 4.

The suggestion is that a true K-alpha x-ray laser be constructed by creating [K] K-shell (or 1s-orbital) vacancies in medium-Z (13–50) atoms through photoionization. Lasing would then take place when a $n = 2$, L-shell electron decays to fill the [K] vacancy (using traditional x-ray terminology appropriate for such deep innershell transitions). Then, to achieve cw-operation, the [L] vacancy created would be filled from outer-shell electrons at a rate exceeding that for total [K] vacancy filling by the laser event.

At first thought this would seem to be impossible, since most [K] vacancies decay by very rapid Auger processes. However, such K-shell Auger transitions create double [LL] vacancies at a different binding energy than for a single [L] vacancy. Hence, the absorbing transition occuring between [LL] and [K] states (e.g., $2p^5$-1s) is at a different energy than the stimulated (laser) decay taking place on a [K]-[L] (e.g., $2p^6$-1s) transition, and cw operation becomes possible. As a result of the multitude of outer electrons present, an Auger cascade occurs in the atom. Because of the required filled outer shells, this scheme is very sensitive to photo- and collisional-ionization of the outer electrons.

4. EXPERIMENTS WITH DIRECT
PHOTOIONIZATION PUMPING

4.1. Photoionization of Neutral Sodium

There has been at least one unsuccessful attempt[11] to construct a 372-Å laser based on photoionization of an innershell 2p electron in neutral sodium, as analyzed above. Sodium vapor was irradiated by a continuum of x-rays from a laser-heated tantalum target. The Nd:YAG driving laser was capable of delivering 20 GW of power in 50 ps pulses. This is a period considered to be short enough to pump the inversion prior to the self-termination time of ~ 400 ps. An estimated gain coefficient of ~ 0.05 cm^{-1} was obtained. This was too low to produce measurable amplification without an efficient cavity.

There are also plans[19,20] to pump sodium with the 206, 209 Å Ne-like selenium lasing lines described in Section 2 of Chapter 3. No coherence is required of this x-ray laser beam, only high brightness. The photoabsorption cross-section peaks in this 200 Å region. A gain product $GL > 15$ and a 1 μJ output are predicted for a 1 mJ Se driving laser beam. Success with selective pumping by spectral lines could pave the way for broadband pumping sources, with which multiple levels may be pumped.

4.2. Photoionization of Neutral Neon

The neutral-neon photoionization scheme described in Section 3.3.2 has also been attempted unsuccessfully[15]. For that experiment, it has been suggested that refraction losses in the medium, due to large density gradients, may have contributed to the lack of gain results.

5. ANALYSIS OF AUGER-DECAY PUMPING

Auger decay takes place when innershell vacancies are created, usually by photoionization, and subsequently filled. The release of energy is to an outer-shell electron. This results in its ionization in a totally radiationless transition.

There have been proposals as well as some successful vuv-laser experiments involving the preferential creation of population inversions following Auger decay. Of course for these processes to take place there must first be an innershell vacancy, usually created by photoionization. In this sense, any desired detailed modeling can be carried out as an Auger-decay branch of the analysis in Section 3.

5.1. Lower-Level Vacancy Production

One of the earliest of such proposals was by McGuire[21] in which Na^+ with an induced K-shell vacancy leads, through subsequent K-LL Auger decay, to preferential creation of a double $n = 2$ lower-state vacancy [LL] in F-like Na^{2+} and maybe higher ions. This can be described by

$$X + h\nu \rightarrow X^+[K] \rightarrow X_u^{2+}[LL],$$ (23a)

in x-ray vacancy notation. The process is also illustrated in Fig. 5. In atomic level notation, this can also be written simply as

$$X + h\nu \rightarrow X^+[1] \rightarrow X_u^{2+}[2, 2],$$ (23b)

without specifying s, p, etc. configurations.

Fractional inversions of a few percent are calculated for $n = 3$ to $n = 2$ transitions near 400 Å wavelength. The potentially damaging effects of high pumping powers and high plasma densities on the medium, and in particular on the $n = 3$ population, are recognized. Notice that this is a self-terminating scheme.

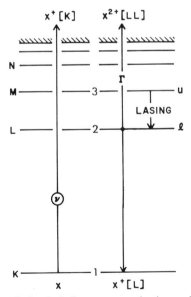

Fig. 5.   Energy-level shematic for L-shell vacancy production and population inversion by Auger decay, which fills a K-shell hole created by photoionization.

5.2. Upper- and Lower-Level Population and Inversion Production

More recent proposals involve additionally the preferential population of the upper-laser level as the result of an Auger processes[22–24]. This concept is similar to that above. It is based on the following reactions, shown for neutral Xe ($4d^{10}5s^25p^6$ ground state) in atomic-level vacancy notation:

$$X + h\nu \rightarrow X^+[4d] \rightarrow X_u^{2+}[5s, 5s], \qquad (24a)$$

and

$$X + h\nu \rightarrow X^+[4d] \rightarrow X_u^{2+}[5s, 5p]. \qquad (24b)$$

These are for populating the upper- and lower-laser levels, respectively. Population inversion results from a larger Auger rate for the first reaction than that for the second. Vacuum-uv lasing at 1089 Å then proceeds by decay of an outer 5p electron to fill the second [5s] vacancy shown for the upper level. This also is a self-terminating scheme.

5.3. Upper-Level Auger Shifting for CW Operation

The reader is also reminded that Auger effects can be used to alter rapidly the energies of upper levels that are vacated by stimulated emission. This can result in shifted absorption transitions and hence possibly cw operation. This is described as a photoionization-pumped cw scheme in Section 3.3.4, rather than an Auger pumping method.

6. EXPERIMENTS ON AUGER-DECAY PUMPING

6.1. Vacuum-UV Cavity Lasing in Auger-Decay-Pumped Atoms

Lasing at 1089 Å in a cavity pumped by Auger-decay has been successfully demonstrated in experiments with neutral xenon[25]. The general scheme is described in Section 5.2. The xenon was photoionized by a pulse of x-radiation from a tantalum target irradiated with a 55J, 1 ns, 10^{12} W/cm^2 beam from a Nd-doped glass laser. An ASE gain coefficient of 0.8 cm^{-1} was measured by varying the length of the plasma (without mirrors). Approximately one year later another group[26] reported a saturated gain of exp(40) and an output of 20 μJ in a beam of 10 mrad divergence on the same transition. This was accomplished with a modest 3.5-J, 300-ps laser, utilizing traveling-wave pumping.

This work was extended to Zn^{2+} for gain products of $GL = 2.4$, 5.1 and 3.2 at wavelengths of 1270, 1306 and 1319 Å, respectively[27]. It is suggested by the authors that such photoionization and autoionization pumped lasers can be extrapolated to shorter wavelengths. One possible limitation in an isoelectronic extrapolation may be the eventual domination of radiative decay processes over Auger, when Z reaches approximately 14.

6.2. Suggested Auger Effects in a Cu Kα Experiment

The only experiment suggesting innershell lasing on traditional x-ray K-vacancies that could possibly be associated with any Auger-decay transitions involved a laser-heated copper gelatin. Collimated emission on the K_α line at 1.5 Å was reported[28]. The results could not be explained by known processes. When attempts to verify the early results by independent researchers at several laboratories and finally with the original apparatus essentially failed, no further attempts were made. Nevertheless, it remains a part of the archival literature in this field.

REFERENCES

1. W. H. Goldstein, B. L. Whitten, A. U. Hazi and M. H. Chen, *Phys. Rev. A* **36**, 3607 (1987).
2. D. H. Sampson and H. Zhang, *Phys. Rev. A* **36**, 3590 (1987).
3. R. C. Elton, in Methods of Experimental Physics, Plasma Physics, H. R. Griem and R. H. Lovberg, eds., Vol. 4A, p. 148 (Academic Press, New York, 1970).
4. Handbook of Chemistry and Physics, current edition (CRC Press, Cleveland, Ohio).
5. B. L. Whitten, A. U. Hazi, C. J. Keane, R. A. London, B. J. MacGowan, T. W. Phillips, M. D. Rosen and D. A. Whelan, "Interim Report of the $J = 0$ Task Force," Report No. UCID-21152 (Lawrence Livermore National Laboratory, Livermore, California, 1987); also in Proc. Lasers '88 Conference, Lake Tahoe, Nevada, December 1988).
6. T. G. Winter, *Phys. Rev. A* **22**, 930 (1980).
7. M. A. Duguay and P. M. Rentzepis, *Appl. Phys. Letters* **10**, 350 (1967).
8. C. W. Allen, "Astrophysical Quantities," Second Edition, p. 91 (Athlone Press, University of London, London, England, 1964).
9. J. A. Bearden, *Rev. of Modern Physics* **39**, 78 (1967).
10. W. L. Wiese, M. W. Smith and B. M. Miles, "Atomic Transition Probabilities, Vol. II: Sodium through Calcium," Report No. NSRDS-NBS-22 (U.S. Government Printing Office, Washington, DC 20402, 1969).
11. E. J. McGuire and M. A. Duguay, *Appl. Optics* **16**, 83 (1977).
12. H. Mahr and U. Roeder, *Optics Comm.* **10**, 227 (1974).
13. S. A. Mani, H. A. Hyman and J. D. Daugherty, *J. Appl. Phys.* **47**, 3099 (1976).
14. S. E. Harris, *Optics Letters* **5**, 1 (1980).
15. P. L. Hagelstein, *Plasma Physics* **25**, 1345 (1983).
16. H. Lundberg, J. J. Macklin, W. T. Silfvast and O. R. Wood II, *Appl, Phys. Letters* **45**, 335 (1984).

17. P. D. Morley and J. Sugar, *Phys. Rev. A* **38,** 3139 (1988).
18. R. C. Elton, *Applied Optics* **14,** 2243 (1975).
19. W. T. Silfvast, O. R. Wood, II and D. Y. Al-Salameh, in "Short Wavelength Coherent Radiation: Generation and Applications", D. Attwood and J. Bokor, eds., AIP Conf. Proc. No. 147, p. 134 (American Institute of Physics, New York, 1986).
20. J. Trebes, *J. de Physique* (Colloque) **C6,** 309 (1986).
21. E. J. McGuire, *Phys. Rev. Letters* **35,** 844 (1975).
22. R. G. Caro, J. C. Wang, R. W. Falcone, J. F. Young and S. E. Harris, *Appl. Phys. Letters* **42,** 9 (1983).
23. W. T. Silfvast, J. J. Macklin and O. R. Wood II, *Opt. Letters* **8,** 551 (1983).
24. A. J. Mendelsohn and S. E. Harris, *Opt. Letters* **10,** 128 (1985).
25. H. C. Kapteyn, R. W. Lee and R. W. Falcone, *Phys. Rev. Letters* **57,** 2939 (1986) and *Phys. Rev. A* **37,** 2033 (1988).
26. M. H. Sher, J. J. Macklin, J. F. Young and S. E. Harris, *Optics Letters* **12,** 891 (1987).
27. D. J. Walker, C. P. J. Barty, G. Y. Yin, J. F. Young and S. E. Harris, *Optics Letters* **12,** 894 (1987).
28. J. G. Kepros, E. M. Eyring and F. W. Cagle, Jr., *Proc. Nat. Acad. Sci., USA,* **69,** 62 (1972).

Chapter 6 | Alternate Approaches

Throughout this book, considerable emphasis is placed on plasma media for x-ray amplifiers. One reason is because plasmas are unavoidable for the high pumping powers required at x-ray wavelengths and without efficient cavities. Hence, it is not surprising that high temperature laser-produced plasmas are the medium in which success has been achieved so far in terms of useful output.

Three other proposed methods of achieving coherent x-rays are harmonic generation, free-electron lasers (FELs) and gamma-ray lasers ("grasers"). Each is treated, respectively, in Sections 1, 2, and 3. The first of these, namely harmonic generation, originates with visible radiation. As a coherent beam, it may prove useful as a "master oscillator (MO)" to be coupled with a plasma "power amplifiers (PA)" in what has been referred to as a "MOPA" synergism of technologies[1]. This has yet to be implemented in the vacuum region of the spectrum, although it has been demonstrated[2] in the near-uv (3530 Å). Free electron lasers for the vacuum-uv and x-ray region as well as gamma-ray lasers are still in the conceptual stage. Such status is similar to that of plasma amplifiers and harmonic generators in the early 1970s.

Several other imaginative concepts that deserve some continual consideration for future development of compact, tunable and efficient x-ray lasers are summarized in Section 4.

1. HARMONIC GENERATION AND FREQUENCY MIXING

1.1. Introduction

Besides "direct" x-ray gain devices such as described throughout most of this book, certain desirable laser features can be transferred to the short wavelength region by "indirect" nonlinear processes such as the generation of higher harmonics. This is made possible through the principles of nonlinear optics. In particular, the main interest here is in the nonlinear response of certain transmitting materials to electromagnetic waves at optical frequencies. This is analogous to the nonlinear response of circuit elements at audio and microwave frequencies, which results in harmonic distortion, parametric amplification, modulation and rectification. For the vacuum-uv regions, gases, vapors and even plasmas are possible conversion media for high transmission. Also, dielectric breakdown limitations in solids, often leading to destruction, are avoided in vapors.

The net result of interest here is the possible use of such an optical parametric process to produce a coherent, monochromatic and (somewhat) tunable radiation source in the xuv spectral region. This would originate from a powerful infrared, visible or near-uv laser. One might then ask why pursue direct amplification devices? The answer lies in the generally low

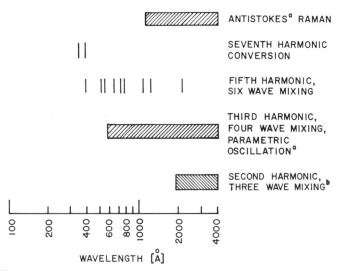

Fig. 1. Wavelength ranges covered by nonlinear optical interactions: (a) tunable over emission band of pump laser; (b) continuously tunable in steps. (Adapted from Ref. 4.)

conversion efficiency in vapors for higher harmonics. Measured efficiencies will be discussed further below.

A particularly enticing goal for such devices is the development of a powerful tunable atomic-hydrogen Lyman-α and/or Lyman-β oscillator/ amplifier at 1216 Å or 1026 Å, respectively. With such a device, neutral hydrogen densities could be measured in thermonuclear-fusion plasma devices by resonance fluorescence. This application is described further in Section 4.2.3 of Chapter 7.

The physical principles involved in such harmonic generation were presented elegantly in an early monograph by Bloembergen[3] in 1965. Since that time the field has virtually exploded, with a host of harmonics in the vacuum-uv region to wavelengths as short as 300 Å. This is illustrated in Fig. 1, taken from a 1985 review article on coherent uv and vacuum-uv sources by Reintjes[4]. Reintjes has also published a recent authoritative monograph covering the broader field of nonlinear parametric processes in liquids and gases[5].

1.2. Nonlinear Polarization

The nonlinear optical processes of present interest involve the elastic interaction of the radiation field with the medium. Conceptually, light that is incident on the medium induces an oscillating dipole moment in the electron or ion distribution. This induced moment or polarization in turn radiates a second optical field that interferes with the incident field. Energy is exchanged among the optical waves, but not with the medium itself. When the incident fields are weak, the potential well for binding the particles is harmonic, the induced polarization depends linearly on the incident field strength, and the induced radiation is at the same frequency as the incident beam. For sufficiently large fields and hence increased displacements of particles from equilibrium, the potential well becomes anharmonic. This results in optical nonlinearities in the polarization.

Mathematically, the polarization ($\mathscr{P}$) induced in a medium by such intense optical fields can be represented by a power series in the total optical field $\mathscr{E}$ as:

$$\mathscr{P} = \varepsilon_0(\chi^{(1)}\mathscr{E} + \chi^{(2)}\mathscr{E}^2 + \chi^{(3)}\mathscr{E}^3 + \cdots). \tag{1}$$

Here ε_0 is the permittivity of free space. The $\chi^{(n)}$s are the nth-order electric dipole susceptibilities, for which the strongest optical interactions occur (compared, e.g., to quadrupole interactions). The polarization becomes a source term in the wave equation, and the nonlinear components give rise to the harmonic frequencies of interest here. The amplitude of the harmonics depends upon the magnitude of the nonlinear (for $n > 1$) susceptibility terms.

Besides the generation of harmonics of the incident radiation, oscillations can also take place at the sum- and difference-frequency combinations when the incident radiation contains components at more than one frequency, i.e., for multi-wave frequency mixing.

1.3. Specific Operating Modes

The two processes that are particularly important for coherent x-ray production are optical harmonic generation and frequency mixing. Another process is four-wave parametric oscillation. In the following, all three of these are categorized and described in more detail for various degrees of nonlinear interaction. Also discussed are the efficiencies achievable and resonant enhancement of the final output.

1.3.1. Second Harmonic Generation and Three-Wave Frequency Mixing

Second harmonic generation at a frequency ω_h can be represented by $\omega_h = 2\omega_i$. The associated three-wave sum- and difference-frequency mixing ($\omega_3 = \omega_1 \pm \omega_2$) involves two different incident fields. Both of these processes result from second-order parametric interactions and are usually the strongest of the various nonlinear orders. Because of certain symmetry restrictions, they are most commonly obtained with anisotropic crystal media.

Second-order nonlinear processes have been used to generate wavelengths as short as 1850 Å, where absorption and dispersion takes over. Besides second-order harmonic generation, three-wave frequency mixing is possible using combinations of two wavelengths, one visible and one infrared, or even one uv and the other visible or infrared. The latter of these combinations has been used to obtain the shortest wavelengths by such a process. Both pulsed and cw conversion to the uv have been achieved. Also, tunable uv radiation has been generated through such interactions, using radiation from pulsed dye lasers either alone or in conjunction with fixed-frequency lasers. Efficiencies as high as 80–85% for production of ultraviolet radiation have been measured.

The wavelengths obtained by second-order interactions are mostly longer than 2000 Å, and hence somewhat longer than those of main interest here. For a more detailed summary of particular experiments, the interested reader is referred to Tables 9–15 in Ref. 4.

1.3.2. Third- (and Higher-) Order Interactions

1.3.2a. HARMONIC GENERATION AND FREQUENCY MIXING. For coherent operation at wavelengths extending into the xuv region, third- and higher-odd-order electric-dipole parametric interactions in transmitting gases and vapors are generally used. (Even-order electric dipole interactions are not allowed by symmetry arguments in such media[4].) Coherent xuv radiation extending from 1000 Å down to 355 Å has been generated on third-, fifth-, and seventh-harmonics ($\omega_h = q\omega_i$, for an order q). The corresponding frequency-mixing interactions are four-wave, six-wave, etc. For example, four-wave mixing takes place with possible combinations of $\omega_4 = 2\omega_1 \pm \omega_2$ and $\omega_4 = \omega_1 \pm \omega_2 \pm \omega_3$. A list of the various nonlinear processes that have been used and the wavelength ranges covered is shown in Table 1, adopted from Ref. 4. In addition, a summary of specific reported xuv results is presented in Table 2, also condensed from Ref. 4. Included in the latter are various xuv wavelengths, pump sources, nonlinear interactions, nonlinear materials and reported efficiencies.

Table 1

Higher-Order Nonlinear Processes Used for UV and VUV Generation[a]

Interaction	Wavelength range [Å]	Driver-laser features[b]
Third-harmonic	570–3550	Tunable
Four-wave mixing	790–2000	Tunable
	1000–2000	Tunable, cw
Fifth-harmonic	386–2110	Tunable, picosecond pulse
Six-wave mixing	590–760	Picosecond
Seventh-harmonic	355, 380	Tunable, picosecond
Ninth-harmonic	1170	Picosecond

[a] Adopted from Table 16 of Ref. 4.
[b] All drivers pulsed, fixed frequency plus additional features noted.

Table 2

Xuv Generation by Frequency Conversion[a]

λ_{xuv} [Å]	Driver	Interaction λ_d [Å]	Material	Efficiency (max.)	Ref.
1064	Nd:YAG	$2 \times 2661 + 5320$	He, Ne	10^{-8}	6
1060[b]	Dye	3×3180	Xe	10^{-6}	7
1026[c]	XeCl	3×3080	Ar, Kr	10^{-8}	8
1026[b]	KrF + Dye	$2 \times 2490 - 5830$	Xe[d]		9
1015–20	Nd:YAG, Dye	$2 \times \lambda_{uv} + \lambda_2$	Xe[d]	10^{-5}	10
730[b]–1010	Dye	$2 \times \lambda_{SH} + $ Dye	Ar[d]		11
972	Dye	$2 \times 2430 + 4860$	H[d]	10^{-9}	12
930	Dye	3×2790	Hg[d]		13
896	Nd:Glass	3×2688	Hg[d]	10^{-6}	14
887	Nd:Yag	3×2661	Ar, H, Ne		6, 15
830[c]	KrF	3×2490	Xe[d], Hg[d], He	10^{-8}	13, 16–18
790[c]	ArF + Dye	$2 \times 1930 + \lambda_{dye}$	H		19
760	Nd:YAG	$4 \times 2661 - 5320$	He, Ne		6
52–748	Nd:YAG + Dye	$3 \times \lambda_{uv}$	Xe[d]	10^{-5}	10
710	Nd:YAG	$4 \times 2661 - 10640$	He, Ne		6
643	ArF	3×1930	H_2[d], D_2[d]		20
			Ar, Kr, CO	10^{-5}	20
626	Nd:YAG	$4 \times 2661 + 10640$	He, Ne		6
570	Xe_2	3×1710	Ar[d]		21
532	Nd:YAG	5×2661	He, Ne, Ar, Kr	10^{-5}	6
498[c]	KrF	5×2490	He, Xe[d]		16
386[c]	ArF	5×1930	He		22
380	Nd:YAG	7×2661	He	10^{-6}	6
355[c]	KrF	7×2490	He	10^{-11}	16

[a] Condensed from Table 22 of Ref. 4.
[b] Tunable over dye laser band.
[c] Tunable over excimer laser band.
[d] Two photon resonance.

The shortest wavelengths, in the 300–400 Å range, have been generated from uv drivers, using harmonics up to 7th order[16,23]. As an extreme demonstration, the ninth harmonic of a 1.06 μm Nd^{3+}:glass laser at 1177 Å has been generated in sodium vapor by a phase-matched 10-photon process[24]. Both fixed-frequency Nd:YAG and tunable (limited) rare-gas halide driver lasers have been used. Dye lasers extend the tunable range. Besides mostly rare gases and molecular gases, there are a few results with metal vapor materials.

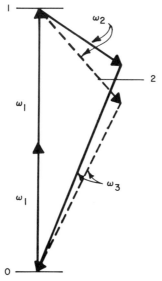

Fig. 2. Energy-level diagram for four-wave parametric oscillation. Pump wave is ω_1 and waves ω_2 and ω_3 are generated. (From Ref. 4.)

1.3.2b. Four-Wave Parametric Oscillation.   This is another third-order interaction and involves the conversion of two photons at a frequency of ω_1 into radiation at two other frequencies ω_2 and ω_3, one higher and one lower, i.e.,

$$2\omega_1 = \omega_2 + \omega_3. \qquad (2)$$

This is accomplished in an atom with a level that is two-photon absorption-resonant with ω_1. It also has a lower-lying level which is close to either the upper or ground level, such that either $\omega_2 \ll \omega_3$ or $\omega_3 \ll \omega_2$. This is illustrated in Fig. 2. This has mostly been accomplished in the vuv region at wavelengths longer than 1000 Å. A specific resonance absorption application involving this process is described in Section 4.2.3 of Chapter 7.

1.3.2c. Anti-Stokes Raman Scattering.   This is a type of two-photon resonant four-wave parametric mixing of the form $\omega_{AS} = 2\omega_1 - \omega_2$, in which the intermediate state has a Raman active transition resonant with the frequency difference $\omega_1 - \omega_2$. Typically, ω_1 is the driver frequency and ω_2 is generated by stimulated Raman scattering. By this process, vuv wavelengths as short as 1380 Å have been generated, not quite in the xuv region of main emphasis here.

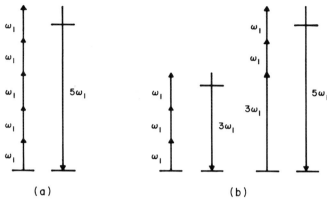

Fig. 3. Energy-level diagram for both direct and indirect fifth-order conversion. (From Ref. 4.)

1.3.3. Combined Processes

Short wavelength radiation can also be produced through multiple lower-order interactions. A fifth-harmonic example is illustrated in Fig. 3. In Fig. 3a is shown the direct five-photon process for generating $5\omega_1$, as discussed above. In a combined process shown in Fig. 3b, the third harmonic is generated first and the output couples with two other photons in a second third-order process, again resulting in $5\omega_1$. The relative conversion efficiencies for the two processes shown depend on the values of the susceptibilities for each as well as the degree of focusing. Hence, either a single-step or a combined process can be favored.

1.3.4. Conversion Efficiencies

The third-harmonic conversion efficiency, given by the ratio of third $W^{(3)}$ to first $W^{(1)}$ harmonic power output is[25]:

$$\frac{W^{(3)}}{W^{(1)}} = \frac{0.082}{\lambda_i^4} N_a^2 [\chi^{(3)}]^2 |\Phi|^2 [W^{(1)}]^2. \tag{3}$$

Here λ_i is the incident wavelength in centimeters, N_a is the number density in the conversion medium in atoms/cm^3, $\chi^{(3)}$ is in esu, and Φ is a dimensionless factor that accounts for focusing and dispersion. When the driver beam is focused and the confocal parameter b is much greater than the length L of the medium, the factor Φ reduces to $(2L/b)^2 \text{sinc}^2(\Delta k_i L/2)$. The wave vector mismatch Δk_i is related to indices of refraction n_3 and n_2 by $\Delta k_i =$

$(6\pi/\lambda)(n_3 - n_2)$. Phase matching with $n_3 = n_2$ permits maximum utilization of the length L. This can be accomplished by an admixture of a substance of opposite dispersion into the medium.

Equation (3), when generalized[6] to any order "q", contains an additional factor $(4W^{(1)}/b\lambda_i)^{q-3}$, which is unity for third-order above. For higher orders, however, the efficiency is an even stronger function of increased power and shorter wavelength. Hence, the efficiency of conversion for high-order harmonics in vapors can exceed that of conversion to lower orders. This has been shown to be by as much as an order of magnitude[6]. This effect is in contrast to the situation in solids, where pumping power is limited by a lower dielectric breakdown (than for gases). Here the conversion efficiency actually decreases with increasing harmonic order.

The efficiency scaling for frequency mixing is similar to that for harmonic generation given above. However, it is more complex and involves each frequency involved. The complete equations can be found in Tables 17 and 18 of Ref. 4.

1.3.5. Resonance Enhancement

Both cw- and pulsed-driving lasers extending from the infrared to ultraviolet have been used as drivers, some even with quite modest powers. Low power operation is possible because of resonant enhancement. This takes place when the sum of partial frequency approximately matches an excitation energy to a discrete bound state. Taking the third harmonic as an example, the susceptibility is given by[26]

$$\chi^{(3)}(3\omega, \omega, \omega, \omega) = \frac{e^4}{\hbar^3} \sum_{i,j,k} A_{ijk} Z_{gi} Z_{ij} Z_{jk} Z_{hg} \quad [\text{esu}]. \tag{4}$$

Here the A_{ijk}'s are frequency-dependent coefficients, with resonant denominators at $(\omega_g - \omega)$, $(\omega_g - 2\omega)$, and $(\omega_g - 3\omega)$ for a fundamental driving laser frequency ω. The frequency ω_g corresponds to transitions to the ground state. The Z_{xy} terms are the dipole matrix elements. By operating near atomic resonances, the susceptibility can be significantly increased by several orders of magnitude, depending on the bandwidth of the pump radiation. An example of this for rubidium vapor is shown in Fig. 4.

Tunable output can be obtained also from resonant enhancement. For example, with two photons used for resonance, a third has no frequency constraint and can be tunable. The most important resonances for frequency conversion processes involve an even number of photons. A widely used

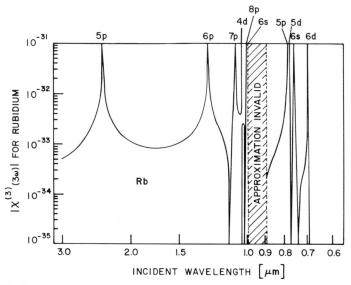

Fig. 4. Nonlinear susceptibility of rubidium as a function of wavelength. (From Ref. 25, © 1973 IEEE.)

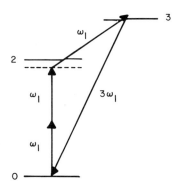

Fig. 5 Energy-level diagram of a two-photon resonance with $2\omega_1$. (From Ref. 5.)

example is the two-photon resonances in a third-order interaction (described above). This is illustrated in Fig. 5.

1.4. X-Ray Parametric Conversion

Parametric conversion consisting of the mixing of optical and x-ray photons in crystalline solids yields the sum and difference frequencies[27,28]. Such techniques can be used to tune the frequency of x-ray lasers efficiently just as they are presently used in the visible spectral region, providing the x-ray intensity is sufficient. The frequency band over which tuning takes place is necessarily small but nevertheless significant and precise.

The process may be described as Bragg scattering from optically induced microscopic charge-stage distributions in the crystal. The induced charge density has a periodicity determined both by the optical wave vector and the crystal. The scattered x-ray beam's frequency shifts up cr down as required by the Doppler process or energy/momentum conservation.

Extensions of this concept include the addition of two visible photons with one x-ray photon, and even the frequency down-conversion of two x-ray beams to produce visible radiation.

1.5. Summary

In summary, harmonic generation and/or frequency mixing can provide coherent xuv radiation with limited tunability, but at a power level considered low for many application. The hope is that it can be further amplified with, perhaps, one of the plasma amplifiers described elsewhere in this book to generate a truly coherent laser beam of high brightness. Tunability of such a combined system will require a broadband or independently tunable amplifier, which represents a further challenge.

2. FREE-ELECTRON LASERS (FELs)

For decades, electrons accelerated in synchrotrons have provided a powerful and useful source of continuum radiation extending into the x-ray region[29]. The radiation comes from "magnetic bremsstrahlung" emission associated with the constantly changing direction, and hence a radial acceleration of the electrons in the transverse magnetic synchrotron field. The radiation is incoherent and is peaked tangentially along the instantaneous velocity vector of the orbiting electrons. It is tangentially extracted and directed into multiple beam ports for simultaneous experiments. In recent years the output power has been enhanced by the addition of high-field oscillating devices called "wigglers" and the more gently oscillating "undulators". Synchrotron sources are inferior in peak brightness to current plasma x-ray lasers such as discussed in Sections 2 of Chapters 3 and 4 by several orders of magnitude[29,30] (see Fig. 4, Chapter 7). However, in terms of average brilliance such lasers would have to operate at a rate of many Hertz to be comparable, instead of in a single-shot mode. Hence the two are complementary as x-ray sources.

Free-electron lasers, sometimes advocated to combine the best of both of these devices, are essentially an outgrowth and extension of modern synchrotron light sources. The goal is to achieve such added properties as coherence, collimation and concentrated power that are characteristic of lasers. They share with synchrotron sources the technologies of low-divergence electron beams and periodic magnetic structures. The major advantages of FELs over conventional lasers is the broad and continuous (over a band) tunability offered. Accompanying these are high power (possible with an electron cloud lasant in vacuum) and high efficiency.

2.1. Basic Principles

The basic FEL concept was presented[31] quantum mechanically by Madey in 1971. Madey also guided the successful efforts to demonstrate the first "optical" FEL amplification[32] and oscillation[33] at an infrared wavelength of 3.4 μm, using the Stanford superconducting rf linear accelerator (linac) as a driver. Classical descriptions[34–36] of the basic FEL principles are applicable for photon energies much less than that of the electrons (typically many MeV) and will suffice here[37].

The typical FEL operates in a linear mode, with a relativistic electron beam from an accelerator traversing a spatially periodic magnetostatic wiggler field as shown in Fig. 6. This wiggler field is normal to the axis of the electron beam. As the electrons traverse the wiggler field, they are driven into transverse oscillations and emit magnetic bremsstrahlung radiation in the forward direction. Hence, the wiggler field simulates the central motion

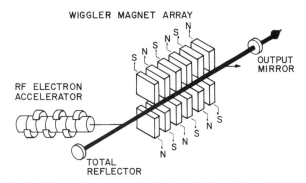

Fig. 6.   Schematic of a free-electron laser oscillator. (From Ref. 43.)

of the synchrotron in an oscillating fashion to maintain an overall linear direction for the electrons as well as the radiation.

2.1.1. Wavelength

In the laboratory frame of reference, the radiation appears at a potential lasing wavelength λ_L of approximately[31,32,37]

$$\lambda_L \approx \frac{\lambda_w}{2\gamma_w^2} = 0.13 \frac{\lambda_w}{E_e^2} \quad \text{cm,} \tag{5}$$

where λ_w is the wiggler spatial period and $\gamma_w = E_e/mc^2$. (Notice that we use λ_L instead of $\lambda_{u\ell}$, so as not to imply specific and bound upper- and lower-laser states as before.) The numerical relation is for an electron energy E_e in MeV. Hence, for electron energies in the MeV range, the wavelength is always considerably less than the wiggler period and decreases rapidly with increasing beam energy. For $\lambda_w = 3$ cm (typical for a magnetic wiggler), $\lambda_L \approx 4$ μm and 3000, 300 and 30 Å for $E_e = 40$, 100, 400 MeV and 1 GeV, respectively.

2.1.2. Bandwidth

Another useful parameter is the bandwidth given by the line-width factor

$$\frac{\Delta\lambda_L}{\lambda_L} \approx \frac{1}{2\mathcal{N}_w}. \tag{6}$$

Here $\mathcal{N}_w$ is the total number of wiggler periods, reasonably chosen to be ≈ 100 times the wiggler length. Hence, $\Delta\lambda_L/\lambda_L \approx 5 \times 10^{-3}$, or about 10 times the typical Doppler line-width factor previously assumed for plasma lasing.

2.1.3. Operation

To understand stimulated emission in an FEL, imagine an optical radiation field (from, e.g., a laser) tuned to near-resonance with λ_L, propagating along the axis of the electron beam, and with an associated transverse electric field (see Fig. 6 again). This additional field can do work on the electrons, causing them either to accelerate or to decelerate depending on the relative phase. They tend to remain trapped by such forces in an energy range corresponding to $\Delta\lambda_L$. In the induced-deceleration case, energy is extracted in quanta from the electron beam to enhance the radiation field. This results in amplification and a fractional energy output corresponding to $\approx \Delta\lambda_L/\lambda_L$, i.e., $1/2\mathcal{N}$ or $\sim 1\%$ of the electron energy. Thus, the beam energy spread required[36] is $\leq 1/2\mathcal{N}$. Through this process, the free electron is stimulated to transfer to a lower (by $h\nu_L$) energy state. The electrons bunch along the axis and may be thought of as radiating from an antenna. To achieve distinct bunching requires a low degree of energy spread, i.e., a high-quality beam. The cooperative nature of the bunches of radiating electrons produces the coherence in the output beam[38].

The trick is to emphasize electron deceleration. This is usually accomplished by adjusting the laser electron beam energy upwards somewhat. Thus, λ_L given by Eq. (5) is slightly less than that of the imposed radiation field, and the electrons are subject to dominant energy loss to the radiation beam. In order to maintain this slightly enhanced electron-energy condition along the length of the wiggler for a fixed λ_L while the beam is losing energy by as much as 10% to the radiation field, the wiggler is sometimes gradually reduced along the length, i.e., "tapered". The purpose is to extract as much energy as possible from the electron beam, with a goal of $> 20\%$, when various forms of energy recovery are included[32,33]. In fact, an increase from 7% to 40% reportedly has been accomplished in the infrared[38-40].

2.1.4. Gain

The approximate net gain per pass GL (following our earlier convention) is given classically as[36,41]

$$GL \approx \pi^2 \frac{\mathscr{I}_b}{\mathscr{I}_A} \frac{\lambda_L^{1/2}}{\lambda_w^{1/2}} \mathcal{N}_w^2 \frac{K_w^2}{(1 + K_w^2)^{3/2}} \approx 10^{-3} \mathscr{I}[A] \mathcal{N}_w^2 \frac{\lambda_L^{1/2}}{\lambda_w^{1/2}}. \tag{7}$$

Here $\mathscr{I}_b$ is the beam current in amperes and $\mathscr{I}_A = ec/r_0 = 1.7 \times 10^4$ amperes is the Alfven current. The wiggler parameter $K_w \equiv eB_w\lambda_w/2\pi mc^2$ contains the wiggler field B_w. This parameter is of order unity for B_w in the kilogauss range and a wiggler spatial periodicity of $\lambda_w = 3$ cm. High gain at a fixed wavelength is best achieved with a large number of short wiggler periods or with high current. For example, at $\lambda_L = 300$ Å and $\lambda_w = 3$ cm a gain product of $GL = 5$ can be achieved with a beam current of ~ 10 A if the total number of wiggler periods is $\mathcal{N}_w = 1000$. While this current is reasonable, the total wiggler length is 30 m, which is extremely long for controlling such a beam. This is the reason that shorter-period optical wigglers with fewer $\mathcal{N}_w$-periods have been proposed, as discussed below.

The gain formula in Eq. (7) can be rewritten in terms of the beam energy parameter g_b using Eq. (5) as

$$GL \approx \frac{\pi^2}{2^{1/2}} \frac{\mathscr{I}_b}{\mathscr{I}_A} \frac{\mathscr{N}_w^2}{g_b} \frac{K_w^2}{(1 + K_w^2)^{3/2}} \approx 7 \times 10^{-4} \, \mathscr{I}_b[A] \frac{\mathscr{N}_w^2}{g_b}. \qquad (8)$$

This illustrates the desirability of less energy for high gain at shorter wavelengths. This is somewhat contradictory to the fundamental requirement for operation at short wavelengths [Eq. (5)]. This means that the high energies required to satisfy Eq. (5) with a magnetic wiggler must be compensated for by increased $\mathscr{N}_w$ and $\mathscr{I}_b$ at shorter wavelengths to maintain high gain. Alternatively, using a very much shortened optical wiggler wavelength requires less energy.

2.2. Current Status

2.2.1. Visible/Ultraviolet FELs

Free-electron laser research is so new and is advancing so fast that the literature is somewhat incomplete. Hence, many results appear first in the popular press and conference proceedings. A recent summary[42] of operating FELs is reproduced in Fig. 7. Ten devices have so far been operated in the μm to visible wavelength range[43]. For infrared wavelengths $\lambda_L \gtrsim 1 \, \mu$m, a linear accelerator or a microtron are suitable drivers. Alternatively, for $\lambda_L \lesssim 0.2 \, \mu$m or 2000 Å, an electron storage ring driver is necessary[36].

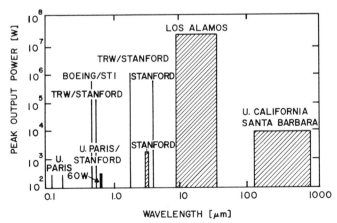

Fig. 7.   April 1988 summary of operating FELs from the far infrared to vacuum-uv harmonics, for wavelengths in micrometers (1 μm = 10,000 Å). (From Ref. 42.)

As seen in Fig. 7, current high-brightness "short wavelength" FEL operation has reached[39,42,44,45] the green spectral region near 5000 Å with an
output approaching a megawatt in a short 2 ps pulse. Coherent emission
can be extended to much shorter wavelengths on harmonics. Such harmonics
are generated by internal nonlinear effects in the wiggler fields as well as in
the electron density[43]. The wavelength achieved is λ_L/n for the nth harmonic,
using Eq. (5). Mainly odd harmonics up to fifth order have been observed
axially in a linear undulator. The radiation is typically three to six orders of
magnitude weaker than the fundamental[43]. For example, a storage ring has
been used to generate FEL amplification in the vacuum-uv region at 1773 Å
and 1064 Å on the third and fifth harmonics, respectively, of an external
(frequency-doubled) 1 μm Nd-Yag laser[39,42,46].

2.2.2. Proposed X-Ray FELs

2.2.2a. MAGNETIC WIGGLER.   With a conventional few-cm-period wiggler,
a beam energy of ≥ 200 MeV is required to reach an xuv wavelength of
≤ 300 Å, as discussed above in relation to Eq. (5). Also, the energy spread of
the beam must be sufficiently small to permit a discrete bunching of the electrons with a periodicity corresponding to the lasing wavelength. The fulfilling
of these criteria generally requires electron storage rings[47,48], rather than a
linear machine such as an electron linac[49] which barely reaches 200 MeV.
Storage rings typically produce circulating beams of well-collimated 500 MeV
electron beams. The electrons can be reused by adding energy to each cycle
(which tends to increase the energy spread of the beam). Alternatively, the
beam energy can be recovered, returned to the accelerating field, and used to
produce a new beam of electrons of limited energy spread. In cavity operation, the amplified radiation field in a resonant cavity is synchronized with
the bunched beam for optimum efficiency. In this mode of operation, less
energy is extracted on each pass and a smaller initial radiation field is needed.
This is a decided advantage in the oscillatory mode of operation (compared
to one-pass amplification).

An alternative proposal[50,51] for a miniature magnetic wiggler would use
sputtering techniques to create thin (1–200 μm) films of magnetic materials
with remnant fields of ~ 10 kG. These films would be used to create an array
of permanent magnets with alternating polarities.

2.2.2b. OPTICAL WIGGLER.   A genuine limitation to achieving efficient amplification at short wavelengths is the practical constraints imposed by λ_w
(and N) for a magnetic wiggler, as mentioned above. One proposed solution

is to replace the magnetic wiggler with a wiggler with a period λ_w in the optical wavelength range. This could provide amplification near 5 Å with a 5 MeV electron beam and a reflecting ring resonator[41]. Then lower energy machines such as Van de Graff accelerators and linacs (both rf-driven for oscillation and the higher peak current induction-driven linacs for amplification) could be used for the x-ray region instead of storage rings.

2.2.2c. PLASMA WIGGLER.   An interesting, recently proposed variation on the optical wiggler would use a series of as many as 96 individual laser-produced plasmas to create[52] a wiggler field of 1 MG or more with an intermediate period of $\lambda_w = 100$ μm. A much shortened wavelength of 0.12 Å or $h\nu = 100$ keV in the gamma-ray region is obtained [Eq. (5)] for a 1 GeV beam energy with this period. The higher field serves to maintain a wiggler factor $K_w \approx 1$. The gain is approximately the same as for a longer wiggler at this beam energy, according to Eqs. (7) and (8) and with $\mathcal{N}_w$ also the same. The quality of the wiggler field is yet to be established.

2.2.2d. SOLID (CRYSTAL) WIGGLER.   There have appeared lately some proposals to use a crystal as a very short period wiggler or undulator. These have been critically evaluated by Fusina[51]. One suggestion is to use a solid-state superlattice, where the beam is directed along a channeling axis at an angle of 45° to the superlattice growth direction[53]. The center of the channeling axis is modulated by the superlattice periodicity, and the particles will be accelerated perpendicular to the direction of motion. A periodicity in the range of 20–200 Å is expected. Hence, the magnetic force in a more conventional wiggler is replaced by the electric force.

Another proposal to use a crystal lattice as a wiggler uses electrons moving in the planar channels of a single crystal[54] (see also Section 4.1.3 on channeling radiation). Here the period would be the same as the channeling wavelength for the electron. It is not clear, however, how the required bunching of the electrons for coherent emission would occur. The predicted x-ray gains with crystal wigglers are low by ASE standards[51]. However, various options exist for improvement, and the miniaturization offered is most intriguing.

2.2.2e. DIELECTRIC CONSTANT IN A STRIATED MEDIUM.   Another proposal to remove the wavelength constraints imposed by magnetic wigglers is to modulate the dielectric constant in a striated medium such as a stacked foil. The plan here is to produce gain in the x-ray regime, e.g., at $\lambda_L = 3$ Å with a 60 MeV driver beam[55].

2.2.3. Cavity Requirements

A basic problem remaining for FEL oscillation on fundamental or harmonics in the xuv and x-ray regions is efficient cavity resonators, as discussed earlier. A typical present FEL gain per pass is in the 10% range. Therefore, efficient cavities (approaching 100% mirror reflectivity) that will also withstand the radiation are essential. Presently a more typical reflectivity in the xuv region is 30% (see Section 2.3 of Chapter 1). Gain products of $GL \approx 1$ per pass are required[36]. Without such high gain, short wavelength operation is probably limited at best to the near vuv region at wavelengths exceeding 1000 Å. Nevertheless, 100 Å operation is optimistically predicted[42,56].

2.3. Summary

Much has been promised and much is expected of FELs in terms of extremely high power and efficiencies in the tens of percent. These expectations are coupled with beam quality approaching the diffraction limit and tunability over a broad range extending from the infrared to the x-ray regions. Continued progress will depend on making good on these promises and the demonstration of usefulness for applications other than as laboratory curiosities[42]. The experiments are extremely costly and time consuming. Much effort has been devoted to extensive theoretical studies, too numerous to have been experimentally verified as yet[57].

3. GAMMA-RAY LASERS ("GRASERS")

3.1. Introduction to the Graser Challenge

It is difficult to imagine a useful x-ray laser operating on an atomic transition at a photon energy exceeding 1 keV (wavelengths < 12.4 Å). For such shorter wavelengths we must turn to nuclear gamma-ray emitting transitions from isomeric states. Here lasing in the 10–100 keV photon energy range (0.12–1.2 Å wavelength) is projected. As mentioned in Section 2.1 of Chapter 1, such wavelengths are actually in the x-ray region by definition. The gamma-ray or "graser" name comes about through the use of nuclear transitions and serves as a convenient distinction of the techniques involved.

A gamma-ray laser operating on a nuclear transition in a solid could have tremendous output power, perhaps as high as 10^{21} Watts[58]. The use of stimulated emission from a metastable isomeric nuclear state for forming a graser was considered as early as 1961[59]. Much of the basic understanding was formulated in the Soviet Union in the 1970s. This and further progress was thoroughly reviewed[60] in 1981. While considerable thought and planning have continued over the years, the present status remains mostly conceptual and at about the level of the x-ray laser a decade or so ago. Certain aspects of the energy storage, inversion and gain mechanisms are considerably more challenging than for x-ray lasers. As one general example, the cross-section for the necessary resonance absorption attains nearly its maximum value in rigid solids, which must also be transparent to the graser beam and still not be overheated by the pump energy required. Nevertheless, the potential payoff in applications is very high. Enthusiastic and active research continues toward the design of a nuclear graser.

In what follows, we will analyze the general requirements for constructing a gamma-ray laser and discuss the present status of attempts to identify suitable media. Much of the basic pumping concepts and analysis developed earlier for atomic/ionic x-ray lasers carries over, as will be apparent.

3.2. Gain Analysis Particular to Grasers

3.2.1. Gain Relations

We will assume for simplicity a two-level system, upper level u and lower level ℓ, where the latter generally consists of nuclei in the ground state and the former of nuclei in the first excited isomeric state. We assume that a nucleus can only be in one of these two states. We further assume that excited-state isomers decay by spontaneous gamma-ray decay into the lower state at a rate $A_{u\ell} = 1/\tau_{u\ell}$, where $\tau_{u\ell}$ is the lifetime of the excited state. Hence, by our earlier definitions, the inversion is self terminating in a time $\tau_{u\ell}$. Hence, the device must be pumped in an interval at least as short as this in order to achieve a population-density inversion. We begin by determining some possible operating parameters.

3.2.1a. THE GAIN PRODUCT.   The ideal gain product for grasers

$$GL = \frac{\lambda_{u\ell}^2}{8\pi} \frac{1}{\tau_{u\ell}} \frac{L}{\Delta\nu_{u\ell}} N_u \tag{9}$$

is similar to that in Eqs. (8) and (10) of Chapter 2, replacing $A_{u\ell}$ by $(\tau_{u\ell})^{-1}$, $\nu_{u\ell}$ by $c/\lambda_{u\ell}$, and taking $F = 1$. Again we can express N_u in terms of the initial state density N_o according to $N_u = N_o(P_{ou}/D_u)$, where D_u (assumed $\approx A_{u\ell}$ here) is the total decay rate of the upper level. We also assume that $P_{ou} = A_{u\ell}$ at least, for self-terminating transitions.

3.2.1b. INTERNAL CONVERSION. More precisely, the total decay rate for nuclear transitions D_u is given by $A_{u\ell} + \Gamma_{ic}$ or $A_{u\ell}(1 + \alpha_{ic})$. Here Γ_{ic} is the internal conversion rate for K-shell electrons in excited isomers. (Internal conversion is the transfer of the excited isomeric-state energy to an inner atomic electron of lower binding energy. As such, it is the nuclear equivalent to the Auger process in atomic systems.) The coefficient $\alpha_{ic} \equiv \Gamma_{ic}/A_{u\ell}$ varies approximately as the ratio $(Z^3/h\nu_{u\ell})^3$. This is not a strong Z-dependence, because $h\nu_{u\ell}$ scales upwards with increasing Z. A value of $\alpha_{ic} \approx 1$ is quite attainable for graser conditions[61]. Notice that this implies $N_u \approx N_o/2 \sim N_o$, in contrast to $N_u/N_o \sim 10^{-3}$ for atomic inversions. This results in far less required pumping power for graser inversion compared to x-rays for transitions of comparable energy. Notice also that the self-terminating time and therefore the graser pulse width, determined by the lifetime $\tau_{u\ell}$, is now shortened to $\tau_{u\ell}/\alpha_{ic} \approx \tau_{u\ell}/2$ when internal conversion is included.

3.2.1c. THE DEBYE-WALLER FACTOR. There is an additional factor which serves to reduce the gain and which can be included at this point. This is the Debye-Waller factor[61] $f_{dw} \approx 0.1$ due to diffuse scattering in the crystal. Hence, Eq. (9) becomes

$$GL = \frac{\lambda_{u\ell}^2}{8\pi} \frac{1}{\tau_{u\ell}} \frac{L}{\Delta\nu_{u\ell}} N_o \frac{f_{dw}}{1 + \alpha_{ic}}$$

$$\approx \frac{\lambda_{u\ell}^2}{160\pi} \frac{1}{\tau_{u\ell}} \frac{L}{\Delta\nu_{u\ell}} N_o. \tag{10}$$

From this, the threshold density for a mirrorless ASE system of gain product GL can be correspondingly written as

$$N_o = \frac{GL}{L} \frac{8\pi}{\lambda_{u\ell}^2} \Delta\nu_{u\ell}\, \tau_{u\ell} \frac{1 + \alpha_{ic}}{f_{dw}}$$

$$\approx \frac{GL}{L} \frac{160\pi}{\lambda_{u\ell}^2} \Delta\nu_{u\ell}\, \tau_{u\ell}, \tag{11}$$

or

$$N_o = GL \frac{8\pi}{\lambda_{u\ell}^3} \frac{\Delta\nu_{u\ell}}{\nu_{u\ell}} \frac{c}{L} \tau_{u\ell} \frac{1 + \alpha_{ic}}{f_{dw}}$$

$$\approx GL \frac{160\pi}{\lambda_{u\ell}^3} \frac{\Delta\nu_{u\ell}}{\nu_{u\ell}} \frac{c}{L} \tau_{u\ell}. \tag{12}$$

Notice that the required density depends directly on the product $\Delta\nu_{u\ell}\, \tau_{u\ell}$. Hence, the wide range of possible nuclear lifetimes can be exploited if there is a flexibility in $\Delta v_{u\ell}$. Otherwise, the density scales up to unreasonably high values for long lifetimes.

3.2.1d. THE LINE WIDTH, LIFETIME CRITERIA.   Using the above values of $\alpha_{ic} = 1$ and $f_{dw} = 0.1$, and assuming $GL = 5$, $L = 3$ cm, and $\lambda_{u\ell} = 0.25$ Å ($hv_{u\ell} = 50$ keV), a density of $N_o \approx 10^{20}$ cm^{-3} is obtained for a product

$$\Delta v_{u\ell} \tau_{u\ell} = 1. \tag{13}$$

This relation is often taken as a convenient criterion at these densities. This density value is already somewhat high by x-ray laser standards, but perhaps not unreasonable for nuclear isomers in solid crystals. Clearly, the product in Eq. (13) cannot exceed unity by more than an order of magnitude or so in a real graser system.

3.2.1e. PUMP POWER LIMITATIONS.   At this point in the analysis a practical restriction enters in the form of the pump power density available. This was discussed earlier (Section 2.5 of Chapter 2) for the x-ray region. From arguments there and recent experience, it is currently reasonable to provide in the laboratory a power density in the lasing medium of

$$\frac{P}{V} = N_u \frac{hv_{u\ell}}{\tau_{u\ell}} \approx 10^{16} \quad \text{W/cm}^3. \tag{14}$$

This limits the lifetime to $\tau_{u\ell} \geq 10^{-10}$ sec, or 100 ps. Lifetimes for E1, M1 dipole-allowed or E2, M2 weakly forbidden quadrupole transitions with a change in angular momentum $\Delta J \leq 2$ in isomers, can be extremely short ($\sim 10^{-17}$–10^{-6} sec). Fortunately, however, isomers can store energy for extended periods in metastable levels and decay over much longer periods on forbidden transitions such as E3, M3 of larger ΔJ. Using Eq. (13), this minimum lifetime (100 psec) leads, for $hv_{u\ell} = 50$ keV ($v_{u\ell} = 1.2 \times 10^{19}$ sec^{-1}), to $\Delta v_{u\ell}/v_{u\ell} \leq 10^{-9}$. Such a line-width factor is very small compared to a value of $\Delta v_{u\ell}/v_{u\ell} = 3 \times 10^{-4}$ used for the x-ray region (which also is typical of normal gamma-ray lines). As we will show, however, operation on such narrow lines may be possible in nuclear isomers for certain Mössbauer transitions.

3.2.2. Possible Losses in the Graser Medium

We can recall from Section 2.2 of Chapter 2 that GL actually represents the *net* gain in the medium. This is the amplification by stimulated emission determined by the cross-section σ_{stim} less the σ_{abs} line absorption. To this line absorption we must now add both the photoionization loss in the medium gauged by σ_{pi}, and perhaps any losses due to Compton scattering (σ_{cs}). We may compare the cross-sections for these reactions to the peak value for stimulated emission, obtained from Eq. (9) according to $G = N_u \sigma_{stim}$ as

$$\sigma_{stim} = \frac{\lambda_{u\ell}^2}{8\pi} \frac{1}{\tau_{u\ell}} \frac{1}{\Delta\nu_{u\ell}} \approx 2.5 \times 10^{-19} \quad cm^2, \tag{15}$$

for $\lambda_{u\ell} = 0.25$ Å ($h\nu_{u\ell} = 50$ keV) and $\Delta\nu_{u\ell} = (\tau_{u\ell})^{-1}$.

3.2.2a. PHOTOIONIZATION.   The above cross-section can be compared with that for photoionization in heavy atoms ($Z \approx 50$), given by

$$\sigma_{pe} \approx 10^{-14} \frac{Z^4}{(h\nu_{u\ell})^3} \approx 5 \times 10^{-22} \quad cm^2, \tag{16}$$

when $h\nu_{u\ell} = 50$ keV is inserted. Hence, photoionization loss of the beam is not a severe problem, at least for the $h\nu_{u\ell} > 10$ keV range considered here and for $\Delta\nu_{u\ell} \tau_{u\ell} = 1$.

Photoionization is a problem with operation on higher $\Delta\nu_{u\ell} \tau_{u\ell}$ products, with even lower $h\nu_{u\ell}$, or with very long lengths for high gain. When this occurs, a possible recourse is to use crystals perfect enough to exhibit the Borrmann effect[62]. Here, photoionization losses in the Bragg direction may be greatly reduced for one of the wave modes which form in a single crystal as a result of Bragg diffraction. As with the Mössbauer effect, heating and other lattice-distortion effects described below must be minimized to achieve the full advantage of this process.

3.2.2b. COMPTON SCATTERING.   When photoelectric absorption is negligible, Compton scattering will remain as the main source of opacity to the graser beam. It can be estimated by[60]

$$\sigma_{cs} = 0.658 Z e^{-0.00295 h\phi_{u\ell}} = 29 \text{ barns} \doteq 2.9 \times 10^{-23} \quad cm^2, \tag{17}$$

for $Z = 50$ and inserting $h\nu_{u\ell} = 50$ keV. While this is small, it can be important in extreme cases (discussed in Section 3.3).

3.2.2c. DIFFRACTION.   A further potential loss to the beam is through diffraction, particularly when the transverse dimension d is made very small. Minimizing diffraction requires that

$$d^2 \geq \lambda_{u\ell} L. \tag{18}$$

This means that the diameter d of a $L = 1$ cm long plasma lasing at a photon energy of 50 keV (0.25 Å wavelength) must be greater than 0.5 μm or 5000 Å. Even for the lower limit, i.e., $h\nu_{u\ell} = 10$ keV (1.2 Å) considered here, a diameter $d \geq 1$ μm is tolerable. These are practical sizes of whiskers or fibers which are proposed as graser media, also chosen to reduce the amount of heat absorbed.

3.3. Operating Modes

3.3.1. General Considerations

3.3.1a. MÖSSBAUER LINE NARROWING.   It was recognized quite early[63] that meeting the above criteria for laboratory graser operation would require some degree of enhanced line narrowing. Such narrowing is possible because most of the line broadening in a pure sample is due to recoil. Recoilless emission can occur in a host crystal by impurity nuclei, according to the Mössbauer effect. The crystal serves to absorb the recoil momentum, thereby reducing the recoil energy from the gamma emission to negligible value. This makes possible, in principle, lines as narrow as the natural width. In fact, line broadening factors of $\Delta\nu_{u\ell}/\nu_{u\ell} = 10^{-15}$ are reportedly measured. Furthermore, values of 10^{-23}–10^{-22} have been predicted[61]. The previous value (10^{-15}) is probably more realistic for a graser environment, which may include spurious broadening[64]. Such high resolution measurements are very difficult to obtain and hence data are sparse.

To achieve maximum Mössbauer benefit, the graser sample has to be pure and imbedded in a carefully prepared pure crystal lattice of uniform temperature (near absolute zero) and field. The crystal quality is very important to minimize additional line broadening due to lattice irregularities and perturbations to the isotropy of the environment by field gradients. Notice that this is a delicate operating condition and pumping must be "gentle" so as not to heat the crystal, etc. Even acoustical vibrations can disturb the narrow resonance achieved.

Therefore, the mode of pumping chosen depends primarily on just how narrow a line can actually be produced in the intense pumping environment needed for graser operation. It is likewise important to ascertain what compromises or parameter trade-offs might be necessary. This requirement on line width translates through Eq. (13) into a choice for the spontaneous emission lifetime $\tau_{u\ell}$ of the upper level.

3.3.1b. EXCESSIVE HEATING. The other major design consideration is the amount of energy that is deposited as heat in the crystal as a result of a particular pumping method. Besides possibly destroying the Mössbauer effect as well as the Borrmann effect for enhanced crystal transmission (discussed further below), this heat load relates to the overall efficiency of the pumping mechanism. The degree of heating is dependent upon the transparency of the crystal and isomer to the pumping energy. Hence, the crystal is often made extremely thin ($\leq 10 \ \mu$m).

3.3.1c. STEPWISE PROCEDURES. Mostly because of the heating problem, virtually all of the proposed schemes assume that excited isomers will first be pumped relatively slowly and gently into extremely metastable states over periods of seconds to hours. This might be accomplished, for example, by irradiation in a high-flux nuclear reactor or charged-particle accelerator until saturation quantities of the isomeric nuclei are obtained. Beyond this the approaches and times involved vary.

Generally, the next step involves the actual physical segregation of the excited isomeric states from other isotopes and nuclei of the same A and Z but different excitation levels. This is necessary because excited states of a particular spin cannot be selectively formed. This separation process would be followed by the condensation or crystallization of the excited isomeric nuclei in a recoilless lattice. This could be a long and slender (diameter/length $= 10^{-3}$–10^{-4}) thread of beryllium (low-Z and hence less absorbing for the graser radiation), perhaps in the form of whiskers, and cooled to liquid helium temperatures[65]. From there lasing is initiated, either directly from the existing state or with additional pumping. The various schemes differ most in the final step here, several possibilities of which will now be discussed.

3.3.2. Specific Graser Schemes

It is convenient to categorize various proposed schemes for graser operation according to their upper-state lifetimes:

3.3.2a. SLOW-TRANSITION SCHEMES. For the very slow mode, i.e., for decay times τ_{ul} measured in minutes, sufficient Mössbauer narrowing ($\Delta v_{ul}/v_{ul} = 10^{-22}$) to satisfy the criterion in Eq. (13) appears very unlikely. Hence, a very high isomer density and/or pumping flux must be used[66]. The obvious inconvenience of such long times has led to suggestions for very slow pumping of the isomers to a metastable inverted state. They are then transferred to the host lattice. Finally, lasing is initiated by a change in environment. The latter may be accomplished, for example, by rapid cooling to reduce the line width and increase the gain above the losses in the medium.

A novel suggested improvement to increase the concentration of nuclei in the long-time mode was made by Letokhov[61], still based on a simple two-level (u and ℓ) atomic model. His method would operate at a somewhat shorter decay time ($0.1 > \tau_{u\ell} > 10$ sec) and with enhanced pumping, such that the line-width factors would remain approximately $\Delta v_{u\ell}/v_{u\ell} = 10^{-14}$, i.e., within the established Mössbauer regime. In order to prepare the sample in a shorter time frame, nuclei excited by slow-neutron capture in a thin target would be extracted from the original material by rapidly vaporizing the target with a laser pulse. Then the particular excited nuclei desired would be separated in the plasma formed. First would come selective excitation of a high-lying state by a second laser tuned to the particular isomerically shifted line. This would be followed by photoionization with a third laser. (The similarity to what is now known as laser-isotope separation is apparent.)

Letokhov in the same paper[61] and in a subsequent one[67] also suggested using gamma-ray resonance absorption pumping, instead of neutron capture, i.e., pumping by a (γ, γ') reaction. The cross-section for photon pumping is much larger than for neutron pumping (values are given below). His proposed matches are featured in Table 3. This listing is presented in a form similar to that for photoexcitation-pumped x-ray laser line-match tables in Section 3 of Chapter 3. In this case, even though the pump lines may have widths comparable to the $\Delta v_{u\ell}/v_{u\ell}$ matches shown, the precision of the nuclear level transition energies is insufficient to draw any conclusions as to the best approach to follow. This was pointed out in Ref. 67 and is still a serious limitation. It was also recognized that the long lifetimes $\tau_{u\ell}$ of many of the upper states listed (and hence the narrow absorption lines) makes line-matching between the pump and the narrow absorber transition extremely inefficient. In fact, this is even more so than for atomic transitions (Section 3, Chapter 3). This led to a three-level scheme similar to those shown in Fig. 8. Here, one of the short-lived and hence broad intermediate states is excited by absorption in a similarly broad pumping spectral line and then decays to fill a long-lived upper-laser state.

3.3.2b. PUMPING METHODS FOR SLOW TRANSITIONS.   One goal is to meet the enhanced conditions required to operate on the long ($\tau_{u\ell} > 0.1$ sec) decay times described so far, while hopefully maintaining a degree of Mössbauer

Table 3

Line Matches for Resonance-Absorption Graser Pumping[a]

Pump	Trans.	$h\nu_{pump}$ (keV)[b]	$\Delta\nu/\nu$ $\times 10^4$	Absorber/Lasant	$h\nu_{abs}$ (keV)	τ_{ul} (ns)
Sr	$K\alpha_2$	14.0979	1.5	$_{25}Fe^{57}$	14.1	100
Ga	$K\alpha_1$	9.2517	52	$_{36}Kr^{83}$	9.3	150
As	$K\alpha_2$	10.5080	7.6	$_{55}Cr^{134}$	10.5	100
Ru	$K\beta_1$	21.6568	26	$_{63}Eu^{151}$	21.6	3–4
Rh	$K\beta_1$	22.7236	99	$_{62}Sm^{149}$	22.5	1–10
W	$L\alpha_1$	8.3976	27			
				$_{69}Tm^{169}$	8.42	3.7
Cu^{27+}	2p-1s	8.3917	22			
Fe	$K\alpha_2$	6.3908	144	$_{73}Ta^{181}$	6.3	6800
Kr	$K\alpha_1$	12.649	40	$_{78}Pt^{193}$	12.7	2.2
Al	$K\beta_1$	1.5574	80	$_{80}Hg^{201}$	1.57	—
Hf	$L\alpha_2$	7.8446	32	$_{94}Pu^{239}$	7.87	—

[a] All combinations taken from Ref. 67 except Cu^{27+} from Ref. 68.
[b] All $h\nu_{pump}$ data from Ref. 69 except Cu^{27+} from Ref. 70.

narrowing. For this, it is necessary to increase either the density to supra-solid levels or the pumping power density above that used in Eq. (14), or both.

A combination of these two options has indeed been suggested. Heating by a DT-reaction would be followed by inertial compression to ignite a fusion microexplosion at supra-solid densities[65,66]. (This is similar to methods presently being explored for creating inertial-confinement thermonuclear fusion with compressed micro-pellets.) The vigorous pumping would come from the neutrons released in the fusion reaction. This suggestion has been severely criticized[71] on the basis of excessive absorption and scattering losses in the medium.

The possibility of utilizing the immense energy available from fission has also been studied[66]. The host crystal would be surrounded concentrically with a fission blanket driven to criticality by a series of laser beams from all sides. The beams would be timed so as to create a traveling wave of excitation with the speed of light down the length of the lasant medium. The result is a miniature nuclear explosion, accompanied by the release of copious pumping neutrons[72].

For both of these modes of operation, generally fast neutron excitation pumping has been suggested, i.e., neutron capture in the next lighter isotope. However, the neutron burst must be extremely intense, thereby risking destruction of the crystal lattice before the graser action is complete. This is indeed a most severe problem for gamma-ray laser development of this type. Hence, it was also suggested early that moderated slow (few 10s of eV) neutrons might be used for pumping[65].

Because such enhanced pumping clearly pushes the limit of existing technology, the gains achieved will be marginal, at least at first. Then certain absorption losses enter, as described in Section 3.2.2. The relative magnitudes depend on the particular isomer chosen and the photon energy. Additional effects such as Borrmann-enhanced transmission through a perfect crystal (Section 3.2.2) must be included.

3.3.2c. FAST-TRANSITION SCHEMES.   In the range of $10^{-10} < \tau_{u\ell} < 10^{-5}$ sec, the line width-factor from Eq. (13) is $10^{-14} < \Delta v_{u\ell}/v_{u\ell} < 10^{-9}$ for 50 keV photon energy. Hence, the longest time and narrowest broadening limits are within the range of what is considered reasonable values[64] for Mössbauer narrowing in a graser environment. The lower time limit here is set by pumping power density considerations from the previous section. Obviously, times in the 1 ns to 1 μs range are the most desirable as far as Mössbauer narrowing, reasonable pumping power, and graser duration in a cavity mode of operation are concerned.

For operation within this time-frame there has been a recent rebirth[64] of the gamma-ray pumping idea of Letokhov[61] mentioned above. This approach uses intense incoherent x-ray line emission from, for example, a laser-produced plasma, to pump an excited state. The Breit-Wigner cross-section at 10 keV photon energy for this process is typically 10^{-18} cm^2, compared to $\sim 10^{-24}$ cm^2 for neutron excitation. Again the intermediate excited state that is pumped in a three-level system is selected to have a very short lifetime for maximum broadening, in order to overlap with a pumping line of uncertain width. This storage state then rapidly decays to the upper-laser level whose spontaneous-decay lifetime is at least as long as the driving laser/x-ray pulse.

There are two possible modes of operation, illustrated in Fig. 8. In the first (a), the short-lived isomeric storage state is pumped from the ground state in a high-energy transition. In the second (b), excited isomeric states are preformed over a long period of time by, e.g., thermal neutron capture in a reactor. Next, a relatively nearby short-lived isomeric state is photon-pumped as a temporary storage state, decaying then to the upper-laser state as before. This latter (b) case offers considerably higher (at least 30 to 100 times) quantum efficiency. To reemphasize, the very perplexing problem of finding extremely precise coincidental line-matches in the x-ray region remains a severe limitation in designing an experiment. This is because

226   6. **Alternate Approaches**

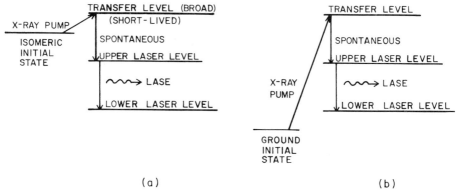

(a)   (b)

Fig. 8. Energy levels and transitions for x-ray pumping of: (a) an isomeric excited state, and (b) a ground state isotope. (Adapted from Ref. 64.)

absorption energies as well as lifetimes for isomeric states are still not available with the required precision.

One particularly promising combination currently studied[68] uses a 8.39 keV He-like Cu^{27+} $n = 2$ to $n = 1$ resonance line from a laser-heated copper-coated target to pump a spin $(3/2)+$ level at 8.4099 keV in Tm^{169}. (In fact, this same isotope was one of those suggested early by Letokhov[67] and is included in Table 3.) Another nuclei which is under current consideration is La^{137}, for grasing at 10.6 keV (for which an x-ray line has not been identified as yet). A further candidate is Ag^{110} for grasing at 117.3 keV, for which there is still insufficient information to complete the analysis.

3.3.3. Raman Scattering Approach

A somewhat novel variation on this laser-pumped graser approach is a non-linear analog of anti-Stokes Raman scattering. Illuminaiion of a nuclear isomer by laser photons in the optical range would generate anti-Stokes Raman gamma-ray output[73,74]. An intense laser would serve to stimulate either an anti-Stokes transition or even a two-photon cascade, in the process of de-exciting a long-lived level prepared by a slow pump. In order to work, it is necessary that the nucleus respond simultaneously to two very distinct frequencies, one optical and the other gamma ray. This may prove a serious obstacle for some time. This is because even the most intense optical fields in present laser focal spots are weak in comparison with nuclear Coulomb fields. It has also been suggested[75] that such transitions might be pumped with a 10 Å x-ray (instead of visible) laser, whenever indeed such a laser becomes practical.

3.3.4. Free-Electron Graser

Finally, a recent proposal is to use a free-electron laser with a plasma wiggler of extremely high magnetic fields to make a gamma-ray laser[52]. This is somewhat similar to proposals for using wigglers of optical wavelength spacings[41]. Both of these are discussed at greater length in Section 2 of this chapter.

3.4. Summary

In conclusion, research on gamma-ray lasers is currently at the stage of x-ray lasers approximately ten years ago. The basic challenge continues to be to cope with the contradictory requirements of an intense pump and a relatively undisturbed solid host material. Efforts are presently aimed at laying the groundwork for future experiments rather than trying to build an actual laser. The consensus appears to be that much additional basic study of the structure and properties of nuclear excitation must precede a true estimate of the feasibility of a gamma-ray laser. Some experts assert that a reasonable time scale for operation of a graser is a couple of decades, i.e., well into the next century[73].

4. OTHER APPROACHES

There is a continual need for different ideas and approaches, perhaps varying radically from those understood to date and described in this book. That may indeed provide the impetus needed to progress to improved efficiency for higher output, to beam quality, and to tunability. Some possibilities that have been proposed are described briefly below. Most are quite speculative, with little practical gain analyses. In contrast to the advanced schemes discussed in more detail in the prior three sections, these concepts are not being extensively pursued at present. Nevertheless, they should not be dismissed completely at this point.

4.1. X-Ray Lasing in Crystals

X-ray lasing in a solid substance such as a crystal is a most intriguing challenge. Two obvious advantages are the high densities and the low temperatures available, which combine to give high gain in a small volume. Coupling this with lasing on core K-shell transitions would lead to high overall efficiency. These are lofty goals, but worthy of consideration.

As a forerunner to x-ray lasing in crystals, some work has been done on measuring luminescence excited by synchrotron radiation[76]. Vacuum-uv luminescent peaks at 6.7 eV and 6.8 eV (1850 Å and 1823 Å) were identified in BeO and MgO, respectively.

4.1.1. Internal Fluorescence Pumping

It has been suggested[77,78] that photon pumping might take place locally in a crystal doped with a lasant material by the absorption of localized bremsstrahlung fluorescent radiation. This would provide a bulk source of continuum radiation at homogeneously distributed emission centers. For example, a perfect crystal of beryllium could be doped with a concentration of lasing atoms such as nickel or copper. The beryllium continuum emission would then pump the heavy metal. Such high-Z elements were chosen to have a high line-radiation fluorescent yield, narrow lines, and short wavelengths—all favorable for high gain.

On the other hand, beryllium was selected as the lowest-Z crystal available to reduce the fluorescent yield for line radiation and to maximize the continuum emission. Excitation of the continuum fluorescence in beryllium would take place with a pulse of fast electrons. It is also proposed that the crystal lattice serve as the resonant cavity. The development of nearly perfect synthetic crystals by layered synthetic microstructure techniques, discussed in Section 2.3.2 of Chapter 1, may further crystal approaches such as these.

4.1.2. Anomalous Transmission by Diffraction Effects

One major problem with efficient laser generation in a crystal is absorption of the x-rays in the solid material. It was suggested[77] quite early that certain diffraction effects in crystals may effectively offset some of this absorption. One is the Kossel effect whereby interference between waves spreading from individual fluorescing centers within a crystal give rise to a manifold of coherent and collimated beams. Crystals with lower symmetry yield fewer beams, which is desirable for higher output. Another phenomenon that might prove valuable is the Borrmann effect, in which anomalous propagation takes place through crystals in certain directions. This is discussed in Section 3 in regard to gamma-ray laser possibilities.

Das Gupta has claimed[79,80] to have observed the Kossel/Borrmann diffraction enhancements of x-ray transmission in copper, excited by a microfocus x-ray beam. He has associated his results with x-ray gain in a crystal[81]. His reported experimental results include a nonlinear increase in x-ray intensity with excitation current, along with significant line narrowing on K-α lines. He also reported[80] nondivergent discrete x-ray emission at discrete frequencies from cylindrical-bore targets excited by a Van de Graff accelerator. His nonlinear intensity results have also been cited[82] as possible evidence of what is referred to as quasi-Cerenkov radiation due to periodic inhomogeneities in the refractive index in a crystal.

Das Gupta's experimental results and explanation based on a model including parametric coupling between photons and electrons have been carefully scrutinized experimentally[83,84]. The conclusion was that the results which were associated with stimulated emission can be explained on the basis of instrumental effects. Additionally, the discrete frequencies reported by Das Gupta were not observed in these experiments. Therefore, any evidence of x-ray lasers in crystals remains unverified at this time.

4.1.3. Channeling Radiation in Crystals

Radiation is known to result from the interaction between the field in an aligned crystal and charged particles channeled along specific directions[85]. The radiation resulting from the oscillatory motion of the particles may be shifted from a few eV in photon energy to the keV or even MeV region. This occurs as a result of Doppler shifts at relativistic energies[51,85–88]. It has been suggested that x-ray lasing may be induced for channel radiation under certain circumstances[51,88,89].

Two classes of proposed methods have been identified[51]. One approach considers the field as a very short-period undulator for a solid-state free-electron laser. This is discussed in Section 2.2.2d. The second is of a more conventional laser type, and involves transitions between channeling states. The latter is the subject of this section.

The inverted states are the quantum states for the particles which are channeled along a preferred direction in the crystal. These states correspond to different values of the transverse energy. The gain is given by[51,87,89]

$$G = \frac{3}{\gamma_{rel}^2} \frac{\lambda_{u\ell}^2}{\Delta v_{u\ell}} A_{u\ell} N_u F \quad cm^{-1}, \tag{19}$$

where $\gamma = (1 - v^2/c^2)^{-1/2}$ is the relativistic Lorentz factor and $\Delta v_{u\ell}$ is the width of the channeling levels. This is equivalent to the gain formula in Eq. (10) of Section 2.2.2, Chapter 2, for $\gamma = (24\pi)^{1/2} = 8.7$ or $v/c = 0.993$, i.e., for a highly relativistic beam. However, velocities v for ASE-level gains are not yet available. More realistically, at present, a beam current density of 10^6 A/cm^2 and $A_{u\ell}/\Delta v_{u\ell} \approx 1$ results[51] in $G = 2.1 \times 10^{-3}$ cm^{-1} for a laser wavelength of $\lambda_{u\ell} = 10$ Å.

As with gamma-ray lasers (Section 3), ultilization of the Borrmann effect to enhance the transmission of x-rays through the crystal has been projected to increase the available gain in the above example by a factor of 100. This was proposed in Ref. 89 and debated in Ref. 51. A novel curved-crystal design[89] consisting of a series of bent crystals forming a complete circle would essentially form a ring laser and eliminate the need for a reflecting cavity. Distributed feedback (Section 2.3.2 of Chapter 2) also could prove useful for enhancing the output and reducing the beam requirements.

4.2. Stimulated Transitions Involving Free Electrons

4.2.1. Stimulated Bremsstrahlung

The concept of making a maser based on induced transitions of free electrons in the Coulomb field of a nucleus apparently was advanced first by Marcuse[90] in 1964. A laser based on this concept would in principle be tunable, because of the spread in energy for electrons in non-bound states. Marcuse's suggestion was based on free-free transitions, leading to stimulated bremsstrahlung emission. The concept is not limited to Coulomb fields, and in principle will work with any other electric or magnetic field. As such, this was likely a forerunner of the free electron laser (FEL) approach described in Section 2. The densities required even for microwave amplification are only available in crystals. Even without lasing, this process could serve to additionally cool free electrons and perhaps enhance collisional-recombination pumping, discussed in Chapter 4.

4.2.2. Stimulated Recombination

Stimulated recombination of free electrons into bound states in a Coulomb field was analyzed by Altshuler[91] in 1973. He mainly studied the added effect on populations and ionization balance (see Section 3.2.2a of Chapter 2). One conclusion was that if it is possible to measurably perturb the population distribution of an atom with a high power laser, it may be possible to deduce the free-electron velocity distribution from the cascade spectrum. Cross section estimates indicate that it is not possible to alter the population densities of bound states (compared to equilibrium values) sufficiently to create significant population inversions. The central problem is a lack of sufficient electron density in a specific free-electron energy interval. Short of direct pumping of gain, this process could serve to enhance collisional recombination into excited states, as discussed in Chapter 4.

More recently, Fill[92] restudied this stimulated-recombination concept, emphasizing the tunable short-wavelength laser applicability. Fill assumed the existence of a non-equilibrium plasmas and analyzed the possibility of lasing directly into the ground state. This would give a high photon energy in a very short time in a self-terminating mode. As an example, for recombination onto protons leading finally to the $n = 1$ state of hydrogen, a gain coefficient of 2.3 cm^{-1} at a wavelength of 912 Å was predicted for a proton density of 10^{20} cm^{-3}. A scaling of $\sim Z^2$ for the density was indicated. This could be important for reaching shorter wavelengths with higher-Z hydrogenic ions. However, it was not analyzed in detail.

Somehow, the electron density per energy interval must be increased in order for these processes to be significant for x-ray lasing. It is possible to identify three possible scenarios to achieve this goal:

(a) a non-Maxwellian velocity distribution for the free electrons,
(b) a narrow velocity distribution around a specific value, such as found in a monoenergetic electron beam, and
(c) in addition to (b), a high degree of directionality for the beam.

Of the three, (a) is considered the most likely to be experimentally verified[92].

Pumping requirements and competing processes for these free-electron schemes have not been addressed. This remains a crucial question as to the viability of this scheme for creating significant gain in the xuv spectral region. All in all, this remains a concept that has serious problems in becoming a practical candidate for x-ray lasing, unless a truly non-equilibrium situation can be generated at high particle densities. This of course would enhance a number of other candidates. Nevertheless, the potential for tunability remains a very attractive feature of the stimulated recombination concept for future applications.

4.3. Stimulated Compton Scattering

In stimulated Compton scattering, photons injected into an oncoming relativistic electron beam are scattered back at greatly increased energies[93,94]. The additional energy derives from the electron beam. Hence, it may be possible to reach well into the x-ray region, starting with a long-wavelength laser. In addition, the output wavelength would be tunable over wide frequency ranges by varying the acceleration potential in the electron beam [see Eq. (20)].

For a colinear geometry with an electron beam of energy E_e pointed towards a photon beam of frequency v_1, the frequency v_2 of the backscattered wave is given by the relativistic Doppler formula

$$v_2 = 4v_1 \left(\frac{E_e}{E_0}\right)^2, \tag{20}$$

where $E_0 = 0.5$ MeV is the electron rest mass. For 5 MeV electrons, 2000-Å photons then emerge as 5-Å x-rays. With accelerators capable of exceeding 1 GeV, even shorter wavelengths are conceivable.

As with other lasers, for stimulated scattering one would expect an exponential growth in intensity with time or distance. The gain coefficient is given by[94]

$$G = 0.7 r_0^2 \frac{E_e h v_2}{(\Delta E)^2} \lambda_1 \lambda_2^2 N_v N_e \tag{21}$$

where λ_1 and λ_2 are again the incoming and emitted wavelengths, N_v is the photon density in the incident beam, and ΔE is the linewidth of the energy scatter of the electrons. Hence, for an electron-beam voltage of 2 MeV, an electron energy resolution of $\Delta E/E = 10^{-5}$, a beam electron density of $N_e = 2 \times 10^{13}$ cm^{-3} (corresponding to a current density of $\sim 10^5$ Å/cm^2), and a photon density of $N_v = 1.8 \times 10^{22}$ cm^{-3}, a significant gain coefficient value of 2.2 cm^{-1} is derived. This is for an initial wavelength of $\lambda_1 = 1.06$ μm (Nd-glass laser) and a final wavelength of $\lambda_2 = 166$ Å. For shorter wavelengths at higher beam energies, an increase in electron or photon densities must be obtained for the same gain coefficient. This is because of the λ_2^2 dependence in Eq. (21).

4.4. Optical Pumping of Relativistic Ions

It has been proposed that the electron beam used in Compton scattering as described above be replaced by a relativistic beam of multiply ionized atoms[95,96]. The population inversion created by optical pumping could emit with tunability in the x-ray region because of the Doppler shift. ASE along the ion column would be detected with an energy of

$$E_{lab} = h v \gamma_{rel} (1 + \beta_{rel}), \tag{22}$$

in the laboratory frame of reference, where γ_{rel} and β_{rel} are the usual relativistic parameters.

An example given in Ref. 95 is a Na-like K^{8+} ion lasing on a 3d $^2D_{3/2}$-3p $^2P_{1/2}$ transition. The normal wavelength[70] of emission is 459 Å. However, in an ion beam of laboratory kinetic energy equal to 10 GeV/nucleon, it is projected that the laser emission will be shifted to 19.8 Å. Also, a gain coefficient of 1×10^{-3} cm^{-1} is calculated. For a 5-m length, the gain product would be $GL = 0.5$ and $\exp(GL) = 1.7$, in the simplest form. Such ion energies in the multi-GeV-per-nucleon range are barely achievable at present.

REFERENCES

1. R. A. Andrews, J. Reintjes, R. C. Eckardt, R. H. Dixon, R. W. Waynant, T. N. Lee, L. J. Palumbo, R. Lehmberg, J. DeRosa and W. Jones, "ARPA/NRL X-ray Laser Program Technical Report," Memorandum Report No. 3130 (U.S. Naval Research Laboratory, Washington, DC, 1975).
2. I. V. Tomov, R. Fedosejevs, M. C. Richardson, W. J. Sarjeant, A. J. Alcock and K. E. Leopold, *Appl. Phys. Letters* **30**, 146 (1977).
3. N. Bloembergen, "Nonlinear Optics" (W. A. Benjamin, Inc., New York, 1965).
4. J. F. Reintjes, "Coherent Ultraviolet and Vacuum Ultraviolet Sources," in Laser Handbook, M. Bass and M. L. Stitch, eds. (North-Holland, New York, 1985).
5. J. F. Reintjes, "Nonlinear Optical Parametric Processes in Liquids and Gases" (Academic Press, New York, 1984).
6. J. F. Reintjes and C. Y. She, *Optics Comm.* **27**, 469 (1978).
7. W. Zapka, D. Cotter and U. Brackmann, *Opt. Comm.* **36**, 79 (1981).
8. J. Reintjes, *Opt. Letters* **4**, 242 (1979); *Appl. Optics* **19**, 3889 (1980); *Phil. Trans. Roy. Soc.* (London) **A298**, 273 (1980).
9. H. Pummer, H. Egger, M. Rothschild and C. K. Rhodes, *J. Opt. Soc. Am.* **70**, 1587 (1980).
10. Y. M. Yiu, K. D. Bopnin and T. J. McIlrath, *Opt. Letters* **7**, 268 (1982).
11. R. Hilbig and R. Wallenstein, AIP Conf. Proc. No. 90 (Laser Techniques for Extreme Ultraviolet Spectroscopy, Boulder, Colorado, 1982), p. 442.
12. B. I. Troshin, V. P. Chebotaev and A. A. Chernenko, *Zh. Eksp. Teor. Fiz. Pis'ma* **27**, 293 (1978) [*JETP Letters* **27**, 273 (1978)].
13. R. R. Freeman, R. M. Jopson and J. Bokor, AIP Conf. Proc. No. 90 (Laser Techniques for Extreme Ultraviolet Spectroscopy, Boulder, Colorado, 1982), p. 422.
14. V. V. Slabko, A. K. Popov and V. F. Lukinykh, *Appl. Phys.* **15**, 239 (1977).
15. S. E. Harris, J. F. Young, A. H. Kung, D. M. Bloom and G. C. Bjorklund, in Laser Applications to Optics and Spectroscopy (S. F. Jacobs, M. Sargent III, J. F. Scott and M. O. Scully, eds.) Physics of Quantum Electronics, Vol. 2, 1975.
16. J. Bokor, P. H. Bucksbaum and R. R. Freeman, *Opt. Letters* **8**, 217 (1983).
17. H. Egger, R. T. Hawkins, J. Bokor, H. Pummer, M. Rothschild and C. K. Rhodes, *Opt. Letters* **5**, 282 (1980).
18. M. Rothschild, H. Egger, R. T. Hawkins, J. Bokor, H. Pummer and C. K. Rhodes, *Phys. Rev. A* **23**, 206 (1981).
19. H. Egger, T. Srinivasan, K. Boyer, H. Pummer and C. K. Rhodes, AIP Conf. Proc. No. 90 (Laser Techniques for Extreme Ultraviolet Spectroscopy, Boulder, Colorado, 1982).
20. H. Pummer, T. S. Srinivasan, H. Egger, K. Boyer, T. S. Luk and C. K. Rhodes, *Opt. Letters* **7**, 93 (1982).

21. M. H. R. Hutchinson, C. C. Ling and D. J. Bradley, *Opt. Comm.* **18**, 203 (1976).
22. H. Egger, T. S. Luk, K. Boyer, D. F. Muller, H. Pummer, T. Srinivasan and C. K. Rhodes, *Appl. Phys. Letters* **41**, 1032 (1982).
23. J. F. Reintjes, C. Y. She, R. C. Eckardt, N. E. Karangelen, R. A. Andrews and R. C. Elton, *Appl. Phys. Letters* **30**, 480 (1977).
24. M. G. Grozeva, D. I. Metchkov, V. M. Mitev, L. I. Pavlov and K. V. Stamenov, *Optics Comm.* **23**, 77 (1977).
25. R. B. Miles and S. E. Harris, *IEEE J. Quantum Electron.* **QE-9**, 470 (1973).
26. J. A. Armstrong, N. Bloembergen, J. Ducuing and P. S. Pershan, *Phys. Rev.* **127**, 1918 (1962).
27. I. Freund and B. F. Levine, *Phys. Rev. Letters* **23**, 854 (1969) and *Phys. Rev. Letters* **25**, 1241 (1970).
28. P. Eisenberger and S. L. McCall, *Phys. Rev. Letters* **26**, 684 (1971) and *Phys. Rev. A* **3**, 1145 (1971).
29. M. J. Weber, in Energy and Technology Review, November-December 1987, p. 1 (Lawrence Livermore National Laboratory, Livermore, California).
30. M. H. Key, *Nature* **316**, 314 (1985).
31. J. M. J. Madey, *J. Appl. Phys.* **42**, 1906 (1971).
32. L. R. Elias, W. M. Fairbank, J. M. J. Madey, H. A. Schwettman and T. I. Smith, *Phys. Rev. Letters* **36**, 717 (1976).
33. D. A. G. Deacon, L. R. Elias, J. M. J. Madey, G. J. Ramian, H. A. Schwettman and T. I. Smith, *Phys. Rev. Letters* **38**, 892 (1977).
34. F. A. Hopf, P. Meystre, M. O. Scully and W. H. Louisell, *Opt. Commun.* **18**, 413 (1976).
35. W. B. Colson, *Phys. Letters* **59A**, 187 (1976).
36. C. Pellegrini, in "Free Electron Lasers," S. Martellucci and A. N. Chester, eds. (Plenum Press, New York, 1980).
37. C. A. Brau, *Laser Focus*, **17**, no. 5, 48 (May 1981).
38. T. Scharlemann, in Energy and Technology Review, December 1986, p. 8 (Lawrence Livermore National Laboratory, Livermore, California).
39. D. R. Bahlman, *Photonics Spectra*, **21**, no. 4, 159 (April, 1987).
40. T. A. Heppenheimer, Popular Science, December 1987, p. 63.
41. P. Dobiasch, P. Meystre and M. O. Scully, *IEEE J. Quant. Electron.* **QE-19**, 1812 (1983).
42. C. A. Brau, *Laser Focus* **24**, n. 4, p. 20 (April 1988).
43. C. A. Brau, *Nucl. Instr. and Methods in Physics Research B* **10/11**, 276 (1985); *IEEE J. Quant. Electron.* **QE-21**, 824 (1985); and *Laser Focus* **23**, no. 2, p. 40 (February 1987).
44. R. Rohatgi, H. A. Schwettman, T. I. Smith and R. L. Swent, Proceedings of the Ninth International Conference on Free-Electron Lasers, Williamsburg, Virginia (1987), published in *Nuclear Instruments and Methods* in *Phys. Res. A* **272**, 32 (1988).
45. J. Adamski, D. Pistoresi and D. R. Shoffstall, ibid.
46. R. Prazeres, J. M. Ortega, C. Bazin, M. Bergher, M. Billardon, M. E. Couprie, H. Fang, M. Velghe and Y. Petroff, ibid, and *Europhysics Letters* **4**, 817 (1987).
47. R. Colella and A. Lucio, *Optics Comm.* **50**, 41 (1984).
48. K.-J. Kim, K. Halbach and D. Attwood, in Laser Techniques in the Extreme Ultraviolet, S. E. Harris and T. B. Lucatorto, eds., *AIP Conference Proc.* **119**, 267 (1984).
49. J. C. Goldstein, B. E. Newnam, R. K. Cooper and J. C. Comly, in Laser Techniques in the Extreme Ultraviolet, S. E. Harris and T. B. Lucatorto, eds., *AIP Conference Proc.* **119**, 293 (1984).
50. F. J. Cadieu, *J. Appl. Phys.* **61**, 4105 (1987).
51. R. Fusina, "Production of Laser Radiation in Aligned Crystal Targets" (to be published).
52. A. Loeb and S. Eliezer, *Phys. Rev. Letters* **56**, 2252 (1986).
53. S. A. Bogacz and J. B. Ketterson, *J. Appl. Phys.* **60**, 177 (1986).
54. V. N. Baier and A. I. Milstein, *Nucl. Inst. and Meth. B* **17**, 25 (1986).
55. M. A. Piestrup and P. F. Finman, *IEEE J. Quant. Electron.* **QE-19**, 357 (1983).

56. Proc. of the Topical Meeting on Free-Electron Laser Applications in the UV, *J. Opt. Soc. Am. B* **6**, 1061–1083 (1989).
57. D. Prosnitz, in Energy and Technology Review, December 1986, p. 3 (Lawrence Livermore National Laboratory, Livermore, California).
58. D. E. Thomsen quoting C. B. Collins, *Science News* **30**, 276 (1986).
59. See R. W. Waynant and R. C. Elton, *Proc. IEEE* **64**, 1059 (1976), discussion beginning on p. 1076 and references 186–189; also see Ref. 60 following.
60. G. C. Baldwin, J. C. Solem and V. I. Gol'danskii, *Rev. Mod. Phys.* **53**, 687 (1981).
61. V. S. Letokhov, *Sov. Phys. JETP* **37**, 787 (1973).
62. B. Borrmann, *Z. Phys.* **42**, 157 (1942); B. W. Betterman and H. Cole, *Rev. Mod. Phys.* **36**, 681 (1964); J. P. Hannon and G. T. Trammell, *Optics Comm.* **15**, 330 (1975).
63. R. V. Khokhlov, *Sov. Phys. JETP Lett.* **15**, 414 (1972).
64. C. B. Collins, et al., *J. Appl. Phys.* **53**, 4645 (1982).
65. V. I. Gol'danskii and Yu. Kagan, *Sov. Phys. JETP* **37**, 49 (1973).
66. L. Wood and G. Chapline, *Nature* (London) **252**, 447 (1974).
67. V. S. Letokhov, *Sov. J. Quant. Electron.* **3**, 360 (1974).
68. C. B. Collins, in *LLE Review* **26**, 61 (University of Rochester, New York, 1986).
69. J. A. Bearden, *Rev. of Modern Physics* **39**, 78 (1967).
70. R. L. Kelly, *J. Phys. and Chem. Ref. Data* **16**, Suppl. No. 1 (American Inst. of Physics, New York, 1987).
71. V. I. Gol'danskii and V. A. Namiot, *Sov. J. Quant. Electron.* **6**, 455 (1976).
72. G. C. Baldwin and R. V. Khokhlov, *Phys. Today* **28**, 33 (1975).
73. G. C. Baldwin and J. C. Solem, Proc. Lasers '81, McLean, Virginia (1981); also Los Alamos Report LA-UR-82-321.
74. J. Hecht, *Lasers and Applications*, April 1987, page 28.
75. E. V. Baklanov and V. P. Chebotaev, *Sov. J. Quant. Electron.* **6**, 345 (1976).
76. V. V. Mikhailin, S. P. Chernov and A. V. Shepelev, *Sov. J. Quantum Electron.* **8**, 998 (1978).
77. L. Gold, *Proc. 1963 Third International Quantum Electronics Conf.*, Paris, France, **2**, 1155 (1964).
78. K. Patek, "Lasers" (SNTL Publ. Co., Praque, 1964).
79. K. Das Gupta, *Phys. Letters* **46A**, 179 (1973).
80. K. Das Gupta, *Phys. Rev. Letters* **33**, 1415 (1974).
81. K. Das Gupta, A. A. Bahgat and P. J. Seibt, *X-ray Spectrometry* **9**, 25 (1980).
82. S. A. Akhmanov and B. A. Grishanin, *JETP Letters* **23**, 516 (1976).
83. H. Aiginger, E. Unfried and P. Wobrauschek, *Phys. Rev. Letters* **35**, 815 (1975).
84. P. J. Ebert and C. E. Dick, *Phys. Rev. Letters* **34**, 1537 (1975).
85. J. U. Andersen, E. Bonderupo and P. H. Pantell, *Ann. Rev. Nucl. Part. Sci.* **33**, 453 (1983).
86. R. W. Terhune and R. H. Pantell, *Appl. Phys. Letters* **30**, 265 (1977).
87. V. V. Beloshitskii and M. A. Kumakhov, *Sov. Phys. JETP* **47**, 652 (1978).
88. M. Strauss, *Phys. Rev. A* **38**, 1358 (1988).
89. Y. H. Ohtsuki, *Nucl. Inst. and Meth. B* **2**, 80 (1984).
90. D. Marcuse, *Proc. 1963 Third International Quantum Electronics Conf.*, Paris, France, **2**, 1161 (1964).
91. S. Altshuler, "The Mechanics of Stimulated Recombination and Conceptual Applications," JILA Report No. 112 (Univ. of Colorado, 1973).
92. E. E. Fill, *Phys. Rev. Letters* **56**, 1687 (1986).
93. R. H. Pantell, G. Soncini and H. E. Putoff, *IEEE J. Quantum. Electron.* **QE-4**, 905 (1968).
94. A. G. Molchanov, *Sov. Phys. Uspo.* **15**, 124 (1972).
95. L. D. Miller, *Optics Comm.* **30**, 87 (1979).
96. G. C. Baldwin and N. J. DiGiacomo, *IEEE Trans. Nuclear Sci.* **NS-30**, 981 (1983).

Chapter 7 | Summary, Applications and Prognosis

1. SUMMARY OF PLASMA LASER PUMPING

In the following section we summarize briefly the promising pump designs described in Chapters 3–5. We later derive from these chapters some design criteria useful in the conceptualization of new x-ray laser schemes.

1.1. Summary of Promising Pumping Modes

Success in achieving meaningful ASE gain and useful output power in the xuv and soft x-ray spectral regions has resulted to date from electron-collisional excitation and recombination pumping in high density plasmas. These and other potentially promising candidates identified so far are summarized in Table 1. Included in this table are columns describing the change Δn in principal quantum number n between the upper and lower laser levels, the

Table 1

Summary of Characteristics of Some Promising Pumping Methods[a]

Method	Δn	No. of Species	Self Regenerative	CW	$G(\lambda_{ul})$ Scaling
e-Excitation	0	1	Y	Y	$\lambda_{ul}^{-1.25}$
e-Recombination	0, 1	2	N	Y	λ_{ul}^{-4}
$h\nu$-Excitation	1	2	Y	Y	λ_{ul}^{-2}
Charge Exchange	1	2	N	Y	$\lambda_{ul}^{-4.25}$
$h\nu$-Ionization	0, 1	2	N	N	$\lambda_{ul}^{-1.8}$

[a] From the analyses in Chapter 3–5.

236

number of plasma species involved, whether the initial atomic state "o" is self-replenished following the pumping and lasing processes, whether the scheme can operate in a cw-mode rather than self-terminating, and finally the scaling with laser wavelength $\lambda_{u\ell}$ of the gain coefficient G. It is seen that all schemes scale in a favorable direction toward shorter wavelength; the rapidity varies, however. Variations and expansions on this table are, it is hoped, possible. We summarize these methods individually following Table 1.

1.1.1. Collisional Excitation

Electron-collisional excitation is the most familiar pumping method, drawing on experience with ion lasers in the near-uv region. It is inherently cw in that the initial and final laser states are the ground level, and that only one ionic species is involved. It also has the advantage that lower-state opacity is often relatively low, so that diameters measured in fractions of a millimeter and lengths in centimeters are reasonable scales. A disadvantage is that it operates on $\Delta n = 0$ transitions. It therefore requires highly charged, high-Z atoms and a large amount of ionization energy. Also, it does not readily scale below $\approx 100-150$ Å with meaningful gain. Another disadvantage is that it requires a relatively high electron density that can lead to refraction loss in the laser beam as it passes through the amplifying medium.

1.1.2. Collisional Recombination

The other approach that has been experimentally very successful is electron-collisional recombination. This method of pumping also only involves one plasma species. However, it does require ions in two successive stages of stripping. Therefore it is not self-replenishing as far as recreating the initial state population as a result of lasing and depletion. The scheme operates best at a low ion temperature. Yet, it requires either a high electron temperature or photoionization to create the initial state. It can operate on $\Delta n = 1$ transitions with atoms in relatively low stages of ionization, requiring less ionization energy from the driver than for the $\Delta n = 0$ scheme described above. This scheme has been successful mostly on hydrogenic and Li-Like isoelectronic sequences. It therefore extrapolates readily to short wavelengths, albeit with rapidly decreasing dimensions because of high opacity on the lower-level depletion transition.

Dielectronic recombination is not really an independent pumping process in terms of achieving meaningful gain. However, it is an auxiliary process that must be accounted for in any modeling that has an initial-state ion with at least one bound electron (i.e., less than completely stripped).

1.1.3. Photoexcitation

The above two pumping methods are now well proven in laboratory experiments. A more complex approach that has proven to be more elusive experimentally is the selective photoexcitation method. This pumping scheme requires precise matching of the pump- and the absorbing-transition energies. Besides this, the hot and dense pumping plasma must be very close to the tenuous and preferably cool (for maximum gain) lasing plasma. The possibility then exists that thermal coupling and pressure can overheat, overionize, and even seriously deform the laser medium. As with electron-collisional pumping, the initial state is self-regenerative following the lasing and depletion transitions. Hence the overall method is essentially cw-capable. The level-selectivity of the pumping is attractive and suggests improved pumping efficiency. However, overall efficiency depends on both of the plasmas and the radiative coupling between them. This scheme also can operate on $\Delta n = 1$ transitions, and hence scales well to short wavelengths.

1.1.4. Charge Transfer

Another level-selective pumping method is by resonance charge transfer. The atomic process has a large cross-section and therefore the potential of operating at reduced density. However, cw operation requires regeneration the neutral-atom donors in addition to the reionization of the lasing ions. A more fundamental difficulty with this scheme is the achievement of sufficient ion/neutral reactions without preionizing the neutral donor material in the vicinity of the high temperature plasma.

1.1.5. Photoionization

Photoionization pumping is a generally self-terminating method and hence not cw by design. It is also not self-regenerative, a feature not really essential for the short times permitted by the gain quenching time. Narrow-band pumping is advantageous to avoid excessive photoionization out of the upper-laser state. Direct photoionization pumping has yet to be proven experimentally in the xuv region. However, Auger pumping following photoionization is currently producing laser action in the vacuum-uv region.

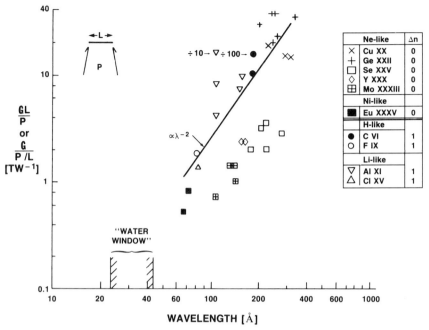

Fig. 1. Survey of measured efficiency as a ratio of gain product (*GL*) divided by input power *P*. The various experiments are described in Sections 2 of Chapters 3 ($\Delta n = 0$) and 4 ($\Delta n = 1$). The Roman numeral spectral numbers correspond to one less ion stage, e.g., Cu XX spectra arises from Cu^{19+} ions. (Adapted from Ref. 1.)

1.2. Efficiency

Efficiency of pumping when defined as the ratio of the measured gain product *GL* to the input driver power *P* (or equivalently as the ratio of the gain coefficient *G* to the power per unit length *P/L*) is plotted in Fig. 1, as adapted from a 1987 publication[1]. In the key, the $\Delta n = 0$ examples are Ne-like 3p-3s and Ni-like 4d-4p results for electron-collisional excitation pumping (Section 2 of Chapter 3). The $\Delta n = 1$ examples shown derive from recombination pumping (Section 2 of Chapter 4). The specific transitions and sources can be identified from the references in the sections noted. There is a clear implication that the experiments on Se, Y, Mo, and Eu carried out at the Lawrence Livermore National Laboratory (LLNL) are less efficient by this definition than others, even on the same transition. This may be a systematic difference or may be a function of the particular manner and driver with which the plasma is created. There are other anomalies (e.g., a low-gain $J = 0$ to $J = 1$ transition) that somewhat parallel this apparent difference, as discussed in Section 2 of Chapter 3.

1.3. Some Design Criteria for New Schemes

It is useful to list some characteristics desirable in conceptualizing new x-ray laser schemes. These have evolved from the preceding analyses. Clearly not all will be obtainable in a particular scheme. They are unnumbered intentionally so as not to imply a particular priority.

- Operation in a long-pulse mode with a cavity:
  — in cw operation,
  — without traveling-wave pumping,
  — with self-replenishment of the initial state.
- Upper-laser-state is relatively stable against
  — collisional depopulation by excitation or ionization,
  — autoionization and Auger decay,
  — photoionization by the laser beam or pump source.
- Lower-laser-state has
  — minimal trapping on the depletion transition (small diameter),
  — energy to ground state $\gg$ laser transition (reduces quenching).
- Low density, to minimize
  — collisional mixing,
  — refraction losses in laser beam.
- High gain:
  — long length,
  — low ion temperature for narrow lines.
- Extrapolates to shorter wavelengths (e.g., isoelectronically), theory accurate and reliable.
- High efficiency:
  — low stage of ionization,
  — efficient pump energy generation and coupling to the lasant.
- Utility:
  — characteristics compatible with existing technology,
  — compact,
  — efficient.

2. PROJECTED X-RAY LASER CHARACTERISTICS

Prior to discussing potential applications, it is useful to project the likely operating parameters of a practical x-ray laser, based on present knowledge. These are summarized in Table 2 and described afterward. Within the antic-

Table 2

Anticipated X-ray Laser Operating Parameters

Parameter	Present	Projected
Configuration	ASE	Limited Cavity
Wavelength [Å]	50–200	5–25
Gain Medium	Plasma	Plasma
Size		
Length [cm]	1–5	10
Breadth [μm]	200	30
Beam Spread [mrad]	10	3
Output Power [MW]	1–10	100
Efficiency	10^{-6}	10^{-3}–10^{-4}
Pulse Length [ns]	0.1–1	0.001–10
Line Width [$\Delta\lambda/\lambda$]	3×10^{-4}	8×10^{-5}
Brightness[a]	10^{30}	10^{33}
Spatial Coherence		
Transverse [μm]	100[b]	500[b]
Longitudinal [μm]	30	120
Temporal Coherence [ps]	1	4

[a] In units of photons/sec-cm²-ster in a $\Delta\lambda_{ul}/\lambda_{ul} = 10^{-4}$ bandwidth.
[b] For a distance of $L = 1$ meter.

ipated ranges for each parameter, a reasonable value is taken for the sake of projection of other parameters.

These estimates are based on current operating lasers. It is possible that entirely different configurations will evolve, with greater efficiency. Devices in stages I and II of evolution (Chapter 1, Section 1.5) such as gamma-ray and free-electron lasers are not included here. At present, it is difficult to project their eventual operating characteristics.

2.1. Configuration

In the currently interesting wavelength range of 10–100 Å, the mode of operation will be predominantly single- or double-pass ASE until efficient cavities become available. Single-transit saturated amplification will occur[1] for $GL \gtrsim 14$, i.e., $\exp(GL) \gtrsim 10^6$ (see also Section 2.2.3 of Chapter 2). Both double- and triple-pass configurations have been tested already in short cavities with sufficiently long pulses (Sections 5.5 of Chapter 2).

2.2. Wavelength

For the near future, wavelengths $\lambda_{u\ell}$ of active x-ray laser research will extend over the 25 to 1000 Å region, with the 25–200 Å range being of most interest as far as applications are concerned.

2.3. Gain Medium

The x-ray lasing medium in most cases will be a highly ionized, medium-Z plasma, which is transparent to soft x-rays. Non-core x-ray transitions likely will be utilized for quasi-cw operation.

2.4. Size

The active medium likely will have a length L of 1–10 cm and a diameter of $d = 0.1$–1 mm, for an aspect ratio of $\approx 100:1$ (see Fig. 2). This may be intentionally altered towards a higher aspect ratio in order to achieve improved transverse spatial coherence, as discussed below.

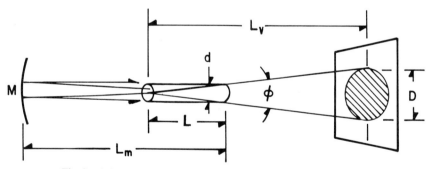

Fig. 2.   Schematic diagram of the ASE amplifier to be characterized.

2.5. Collimation and Beam Spread

The corresponding maximum geometrical beam divergence ϕ for $L = 1$ cm and $d = 100$ μm is then

$$\phi = \frac{d}{L} \approx 10 \text{ mrad.} \tag{1}$$

This corresponds to a beam spread ϕL_v at a $L_v = 1$ m viewing distance, of $D = 1$ cm, or an area of total illumination of 1 cm^2. With the addition of a rear mirror M a distance L_m, L is replaced by L_m in Eq. (1). Then the beam spread is greatly reduced for the second pass, when it is temporally resolved[2]. Intensities comparable to single-pass first-pulse operation can be anticipated in a 3 mrad beam, i.e., with an aperture d reduced from 100 μm to 30 μm.

2.6. Output Power

For the near future at least, readily available pumping powers will probably be in the TW regime. For laser drivers this translates into 1, 10, 100 or 1000 J in 1, 10, 100 ps or 1 ns, respectively. With current efficiencies for conversion from pump to x-ray laser output of $\sim 10^{-6}$, present x-ray laser outputs are in the 1–10 MW range. We hope this will increase to the 100 Megawatt range with improved efficiency in the future. The addition of a rear mirror that captures the rear emission and reamplifies it through the medium tends to increase the total output by approximately two-fold with presently available mirrors[2,3].

2.7. Pulse Length

Pulse lengths will probably range from picoseconds to nanoseconds, the former with self-terminating ASE or traveling wave operation and the latter with multiple-pass cavities.

2.8. Line Width

The line width with significant gain can be anticipated to become $\lesssim 8 \times 10^{-5}$ in terms of $\Delta\lambda_{u\ell}/\lambda_{u\ell}$ (we assumed 3×10^{-4} for Doppler-broadened plasma lines in the analysis in Chapters 3–5). The decrease may be reached by unusual cooling or by gain narrowing as the laser approaches saturation. However, it probably will not begin to approach natural broadening, where $\Delta\lambda_{u\ell}/\lambda_{u\ell} \approx 10^{-7}$ is typical at 100 Å.

2.9. Coherence

2.9.1. Transverse Spatial Coherence

Spatial coherence in a direction transverse to the x-ray laser beam is essential for some applications such as holography, and desirable for others requiring a small focal region. The extent D_{coh} of a region of coherence obtained at a point of illumination of an object at a distance L_v from a source of width d (see Fig. 2) is given from elementary diffraction theory by

$$D_{\text{coh}} \approx \frac{L_v \lambda_{u\ell}}{d}. \tag{2}$$

Hence, a $\lambda_{u\ell} = 100$ Å laser of output aperture $d = 100$ μm will coherently illuminate a $D_{\text{coh}} = 100$-μm object at a distance of $L_v = 1$ m (or a 1-mm object at $L_v = 10$ m, which is a typical source-to-object distance for synchrotrons, as discussed later). Such a coherent source is already of interest in microscopic studies, as described in Section 4.4.1.

From an intensity viewpoint, however, these numbers are less exciting. For example, the area coherently illuminated compared to the total area illuminated is given approximately by the ratio

$$\frac{D_{\text{coh}}^2}{D^2} = \left[\frac{\lambda_{u\ell} L}{d^2} \right]^2, \tag{3}$$

using Eqs. (1) and (2). For $L_v = 1$ m, this is 10^{-4}. The inverse of this indicates that there are 10^4 separate islands of coherence, also known as "transverse modes" in laser jargon. Hence, the power delivered into a single mode by such a device would be only $\approx 100–1000$ W, according to Section 2.6. This also implies that the transverse coherence is restricted to dimensions equivalent to $\sim 1/100$ the beam diameter.

This single-mode power can be greatly increased by reducing the source diameter, in a technique borrowed from high-power infrared laser designs. Equation (3) shows that the power ratio scales inversely as the fourth power of the source diameter, for a given laser length and wavelength. It does not depend on the distance of observation L_v. For example, reducing d from 100 μm to 30 μm by, e.g., reflection from a rear mirror[4], increases the output power by 120 times to the quite useful 10–100 kW range. It also increases the coherence size to $D_{\text{coh}} = 300$ μm at $L = 1$ m, i.e., greater than 10,000 wavelengths. This allows holograms of up to a millimeter in extent to be produced[5].

A further design for achieving a high-brightness, single-mode, spatially coherent x-ray laser, based on these concepts and on existing technology, was described in Ref. 6 and is shown in Fig. 3. Here the x-ray laser beam is reflected from a concave mirror and focused onto a spatial filter to improve the beam quality. (This again is a technique adapted from long-wavelength laser technology.) The emerging beam is redirected by a second concave mirror parallel through a second amplifier. There is enough input intensity

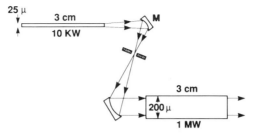

Fig. 3. Possible set-up for using a low-divergence oscillator to drive an amplifier into a fully coherent output beam. (From Ref. 6.)

even after two reflections to dominate spontaneous emission at the amplifier input. This again is the master-oscillator, power-amplifier (MOPA) concept for achieving a high-power coherent beam described earlier (see Index). In this case, however, the oscillator and amplifier are of the same material and wavelength, rather than, e.g., a harmonically generated driving source that must be matched or tuned to the amplifier. This concept could conceivably be improved still further by incorporating a rear mirror in the oscillator, as described in the previous paragraph.

2.9.2. Longitudinal Spatial Coherence

The longitudinal coherence length is given by

$$L_c = \frac{\lambda_{u\ell}^2}{\Delta\lambda_{u\ell}} = 30 \quad \mu m, \tag{4}$$

for a laser wavelength of $\lambda_{u\ell} = 100$ Å and $\Delta\lambda_{u\ell}/\lambda_{u\ell} = 3 \times 10^{-4}$. This is reasonable for imaging biological objects of $>1\ \mu m$ depth. It can be projected to increase by a factor of four for line narrowing, as described above.

2.9.3. Temporal Coherence

Emission from an ASE device maintains coherence for a period t_{coh} approximately equal to the spectral bandwidth, i.e.

$$t_{coh} \approx \frac{1}{\Delta\nu_{u\ell}} = \frac{1}{\nu_{u\ell}}\left(\frac{\Delta\nu_{u\ell}}{\nu_{u\ell}}\right)^{-1} \approx \frac{3 \times 10^4}{\nu_{u\ell}}, \tag{5}$$

at present. This is equivalent to 30,000 cycles or 1 ps at present and 4 ps projected with additional line broadening, at a lasing wavelength of 100 Å.

3.   COMPARISON WITH OTHER XUV SOURCES

For many of the applications described above, sources other than x-ray lasers may be equally suitable for present needs. Indeed, most areas are already being pursued actively with existing sources. Furthermore, they may be currently available either for immediate application or for preliminary studies prior to the availability of a suitable x-ray laser. It is therefore useful to briefly review such alternate sources as conventional x-ray tubes, electron storage rings/synchrotrons, relativistic electron beams, and laser-produced plasmas, and compare the characteristics with those projected in Table 2 for x-ray lasers.

There are numerous ways in which such a comparison of sources can be and has been carried out. For example, a broadband continuum source such as a synchrotron (Fig. 5) may emit copius amounts of total energy on a practically continual basis, but comparatively small amounts of peak inten-

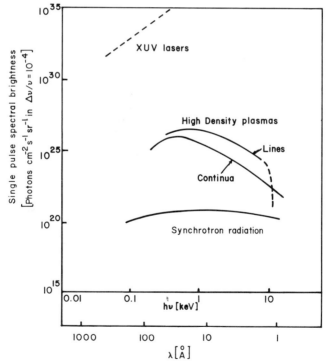

Fig. 4. Comparison of absolute spectral brilliance of various xuv and x-ray sources. (Adapted from Refs. 1 and 4.)

sity or brilliance (sometimes termed brightness) measured in a narrow frequency band and narrow pulse. In such a comparison the x-ray laser wins out over all other sources, as will be explained in Section 3.2 and as plotted in Fig. 4. However, if a comparison were made in terms of average brilliance, over an extended period, x-ray lasers would have to operate at rates of many Hertz to be comparable to synchrotrons. This is not possible at present. Hence, often the choice of best source depends on the application.

For the present comparison we shall use the units for peak, pulsed brilliance included in Table 2, i.e., photons/sec-cm^2-ster in a bandwidth defined by $\Delta\lambda/\lambda = 10^{-4}$. The results of the present section are plotted in Fig. 4, adapted from Refs. 1 and 4, and discussed below.

3.1. Electron-Impact Sources

3.1.1. Conventional X-Ray Tubes

The most familiar and conventional x-ray source is a concentrated region of a target bombarded by an electron beam. Such were the first x-ray sources to be developed and also currently the most widely used. The spectrum consists of a blend of both line and continuum emission. Typically tens of kW of power are dissipated continuously over a target area of 0.1 cm^2. Such anode heat dissipation limits the intensity available for voltages above a few keV. For lower energies, space charge effects are the basic limitation[7]. Only about 1% or less of the incident electron energy is converted to x-rays. The maximum x-ray intensity is therefore limited to a value several orders of magnitude less than the other sources shown in Fig. 4 and hence is not included there[8,9]. For x-ray lithography with a high resolution photoresist, exposure times are prohibitively long with such a source. There is a copius amount of literature on such devices. One convenient reference for the present reader and referenced extensively earlier in this volume is a book by Samson[10].

3.1.2. Relativistic Electron Beam Sources

Copious amounts of x-rays are emitted from pulsed, high power relativistic electron beams (REBs) incident onto solid targets. Such beams operate at the megavolt and megampere levels (see Section 4.2.2 of Chapter 2). They cannot be focused to nearly as small a volume as can laser beams. This is due primarily to Coulomb repulsion between the charged particles. They also cannot be conveniently directed into remote and otherwise inaccessible target locations. However, the efficiency for useful driver energy production is expected to be quite high.

Values for the x-ray yield from such vacuum diodes is not available for publication. However, the x-ray yield from such machines operated in the puffed-gas-filled plasma mode corresponds to the dense-plasma level shown in Fig. 4, as discussed in Section 3.3.2.

3.2. Synchrotron Radiation Sources

3.2.1. The Device

Synchrotron radiation is magnetic bremsstrahlung emission emitted when electrons undergo centripetal acceleration in transversing circular orbits in a magnetic field. It is so named after the machines in which it was first observed. An electron storage ring is typically used for modern synchrotron light sources. In a storage ring, the electrons are kept essentially permanently in a circular orbit, so that synchrotron radiation is emitted continuously. The addition of wigglers and undulators in straight sections of storage rings, as described in Section 2 of Chapter 6 on free-electron lasers, has substantially increased the high photon energy emission and the overall intensity in these devices[11].

There are currently numerous synchrotrons dedicated to xuv and x-ray applications, particularly for materials analysis and alteration. As such, there is an extensive amount of literature on the subject[12]. This is not the place to review this very active field. Rather an example is compared to what is currently considered realistic for a xuv laser for certain characteristics, notably spectral brilliance.

3.2.2. The Spectrum

The spectrum of the emission always has the same shape, as shown in Fig. 5, from Ref. 13. It is a broad continuum at long wavelengths with a peak close to the "critical wavelength" and a rapid decrease to zero at shorter wavelengths. This critical wavelength is given by

$$\lambda_{\text{crit}} = \frac{5.6 r_{\text{e}}}{E_{\text{e}}^3} \tag{6}$$

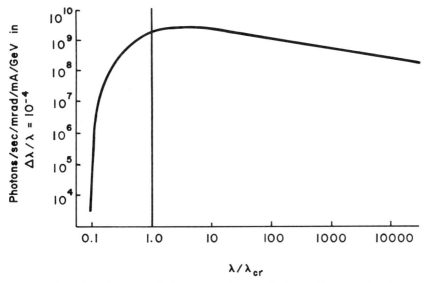

Fig. 5. Universal synchrotron radiation spectral shape applicable to all storage rings. The ordinate units are expressed per beam energy and current in GeV and mA, respectively, and per horizontal beam spread in mrad. For the abscissa, the wavelength is expressed in units of the critical wavelength λ_{crit} defined in Eq. (6). (From Ref. 13)

in Angstrom units, where r_e is the radius of the electron orbit in meters and E_e is in GeV. Hence, a 4 GeV electron beam from a 20-m radius synchrotron emits at a continuum peaked at a wavelength of 1.7 Å. The emission band may extend several orders of magnitude of the high side of the peak wavelength.

3.2.3. The Output of Radiation

3.2.3a. INTENSITY.   The radiation from a storage ring is directed along the tangent to the electron trajectory. Since these tangents completely fill the plane of the circle, the radiation is emitted as a flat horizontal sheet in the plane of the orbit. Ten to twenty exposure stations can be placed around the orbit in a modern synchrotron light source. In the horizontal plane, the angular spread is usually limited by apertures to ≈ 10 mrad. The halfwidth of the vertical spread at the peak wavelength is given in milliradians by $570(mc^2/E_e)$, where E_e is the electron energy[13]. This spread is only 0.1 mrad for an energy of $E_e = 4$ Gev in a direction perpendicular to the plane of the synchrotron orbit. Hence, the solid angle of emission is $\sim 10^{-6}$ steradians.

The emission from a synchrotron source can be accurately predicted, making it an excellent source for absolute intensity calibration in the xuv spectral region. This is not the case for any other source described here. A typical synchrotron might emit $> 10^{14}$ photons/sec-Å in such a solid angle and for a spatial extent of 1 mm^2. For comparison with x-ray lasers, a portion of the radiation is selected by, e.g., a monochromator, of width $\Delta\lambda/\lambda = 10^{-4}$. Therefore, at a wavelength of 100 Å, an intensity exceeding 10^{20} (in the above units) can be expected from presently operating devices. This is plotted in Fig. 4. There is a new generation of synchrotrons machines being planned for an up to 4–5 orders of magnitude increase in brilliance[10], still far lower than expected for x-ray lasers. For comparison with conventional x-ray sources, adequate exposure of a photoresist for lithography by a synchrotron presently requires[7] a period of 10^2–10^4 sec, instead of hours or days.

3.2.3b. POLARIZATION OF THE BEAM. Synchrotron radiation has the additional characteristic that it is highly polarized with the electric vector in the orbital plane. Furthermore, the degree of polarization can be accurately calculated from beam parameters. This is particularly advantageous for xuv applications requiring a polarized beam. One example is the study of anisotropic crystals[8]. Also, in photoelectron spectroscopy of crystal structure, a polarized beam can be advantageous in determining the angular distribution of emitted electrons. Plasma resonances in metals have been related to the polarization properties of the beam. Reflectance measurement with such a polarized beam can be an advantage for determining optical constants of materials in various directions. However, it can also be a handicap when synchrotron radiation is used for efficiency measurements on mirrors and gratings and absolute intensity calibrations. Hence, various orientations relative to the beam must be used to simulate the response to a non-polarized beam.

3.2.3c. BANDWIDTH OF EMISSION. The broadband continuous nature of synchrotron radiation is an important advantage for studying sharp resonances in photoionization continuua and exciton structures in solids. However, ambiguities may enter from scattered light and higher-order contributions in monochromators at a particular wavelength of interest. Selective reflection and filtering as discussed in Section 2 of Chapter 1, as well as double-dispersion spectrometers, are useful. The required high vacuum environment ($< 10^{-7}$ Torr) also can be a problem requiring isolation for some experiments. These are disadvantages that may indeed be overcome with a tunable, narrow-band, intense, collimated x-ray laser, along with a degree of coherence.

3.3. High-Density-Plasma X-Ray Sources

A quite efficient method for converting energy into intense line and continuum x-radiation is a high-temperature plasma. High-temperature plasmas created for this purpose are usually driven either by high-power uv to infrared lasers or electrical discharges. We will discuss each of these approaches.

3.3.1. Laser-Produced Plasmas

High power uv/visible/infrared lasers used for plasma production have the advantage of the concentration of energy into a very small (micrometer-scale) spot on a target. For example, plasma volumes at focus of $\approx 10^{-6}$ cm^3 are typically produced. The target can be at a remote or otherwise inaccessible location, at considerable distance if necessary. The usefulness for line-plasmas for x-ray lasers has already been pointed out at numerous places in this volume. Such laser-drivers result in copius x-ray emission from a near point source of considerable coherence, important for some applications. The useful photon energy can extend from 10 eV to tens of keV. The pulse length can also be very short and flexible, e.g., typically in the 1 ps to tens of ns range. This is very valuable for stop-motion diagnostic applications discussed below. Both line and continuum emission is generated. Efficiencies for production of radiation from certain resonance lines (e.g., He-like $n = 2$ to $n = 1$) can be in the $1-10\%$ range, measured from the driver-laser power. Overall efficiencies for x-ray production have been found to exceed 20%. Some absolute brilliance values are included in Fig. 4.

The brilliance of a laser-produced plasma line source may be obtained from an example where a power of 25 MW was measured in a 11 Å line of width factor $\approx \Delta\lambda/\lambda \sim 10^{-3}$ over 2π steradians[14,15]. The focal volume was $\sim 10^{-5}$ cm^3. This results in a brilliance of $\sim 10^{26}$ in the units defined above. This value agrees with that plotted in Fig. 4 for line emission from a dense plasma.

For materials analysis, laser-plasma x-ray sources offer the advantage of high brilliance, i.e., high energy concentrated in a short pulse and small volume. The combined intense continuum and selected line emissions are a mixed blessing. The difficulties arising from the former have already been discussed in connection with synchrotron radiation. Also, for absorption studies, the multiple line components are sometimes a serious drawback. These are problems expected to be overcome with a monochromatic x-ray laser plasma source.

3.3.2. Electrical-Discharge Plasmas

Pulsed arc plasmas for x-ray production are usually created in high voltage (megavolts) diode devices. Such machines are also used to create the relativistic electron beams described above. Either gases are injected into the diode region or fine wires are exploded. Self pinching to high densities and temperatures follows, along with the copious x-ray emission sought. Such devices for x-ray lasing are also described in Section 4.2.2 of Chapter 2.

The x-ray energy and power obtained are considerably larger than for laser-driven plasmas. For example, a power of 25 GW has been measured from the same 11 Å spectral line used in the example above[16]. However, the size of the electrical-discharge plasma is typically millimeters in diameter and several centimeters in length. Hence the volumes are at least 10^{-2} cm^3 or 10^3 times as large as in the laser-produced plasma example. This results in a comparable brilliance of $\sim 10^{26}$ in the units used above. This is plotted in Fig. 4, where it is included with laser-produced plasmas.

Also, the pulse lengths for discharge plasmas are typically in the 10–100 ns, considerably longer than for laser-produced plasmas. This is an advantage for multipass cavity operation. However, such a long pulse is also a disadvantage for some stop-motion applications. Another disadvantage is the limited access to the source, which is usually imbedded in a large vacuum tank surrounded by insulating materials at the high voltages (MV) present. This becomes a particular problem whenever extensive diagnostics from many directions is essential. Hence, the local environment is quite hostile for some experiments compared to a laser-plasma source which can be placed in any location to which a photon beam can be directed.

4. APPLICATIONS

4.1. Introduction

In most of the papers in the x-ray laser field, there is at most a sentence or two devoted to potential applications. This reflects the general attitude in the scientific community that probably the most important applications will develop after a useful device has arrived. Such applications will depend to a significant extent upon the characteristics of the particular device. For example, a subpicosecond device could open new areas in stop-motion radiography. Likewise, a highly coherent device will offer particular promise for interferometric and holographic applications.

This attitude of invent-it-now, apply-it-later is not without precedent in the laser field, or for that matter in other areas of research. Indeed, it would seem unwise at this point to attempt to narrowly channel research towards very specific applications. This would certainly not be advisable while significant ASE gain is just now being achieved with a few approaches and various basic physical experiments are still being fielded on a rather fundamental scale to explore alternate approaches. Speculation on future applications is likely to be inaccurate. It is often those changes in character that arise from the unforeseen developments that lead to the most significant applications in the long term.

Nevertheless, it is helpful and important to periodically assess the long range applications foreseen in light of current knowledge and needs, albeit somewhat "crystal-ball" in nature. Three serious attempts at such assessments by a host of authorities in various disciplines associated with x-rays have been carried out. The first was organized by the U.S. Naval Research Laboratory and held during a CLEO laser conference[17] in 1973. The second and most thorough of all was compiled and published by Physical Dynamics, Inc. in 1977[8]. The most recent was organized by the Lawrence Livermore National Laboratory[18] in 1985. Meanwhile, there was also a summary included in a 1976 review[19]. Finally, two conference proceedings articles in 1982[20] and 1986[21] emphasized applications. Four specific topics were chosen in the latter, as cited below.

The numerous potential x-ray laser applications described in this section are summarized in Table 3, adapted and updated from Ref. 8. The various areas of application are listed in order of the discussion which follows, rather than prioritized. It is not surprising that many of the predicted applications are of a scientific nature, given the present status of the research in this area. Also, the somewhat unwieldy drivers needed at present lend themselves particularly to research needs.

Indicated in Table 3 are the estimated importances of a short x-ray laser wavelength ($\lambda_{u\ell}$), a high intensity (Int.), narrow lines ($\Delta\lambda_{u\ell}$), good collimation (Coll.), a small beam size, a short pulse length and a high repetition rate, for a particular application. Efficiency is not of major importance at this point. Eventually it may become important for extraterrestrial applications.

The various categories listed in Table 3 are now discussed individually.

Table 3

Summary of Applications

	$\lambda_{u\ell}$ [Å]	Int.	$\Delta\lambda_{u\ell}$ [Å]	Coll.	Beam Size	Pulse Length	Rep. Rate
Scientific:							
Atomic/Accel.	1–200	+	+ +	+ +		+ +	
ESCA	10	+ +	<1 eV[d]		+		
Fusion Diag.	1–1000	+ +	+	+ +	+ +	10^{-12} s	+
Nuclear Decay	5–100	+	+			+	+
Radiation-						10^{-13} s	
Chemistry	1	+ +				10^{-15} s	+
Technical:							
Gratings/Grids		+ +	+ +				+
Lithography	5–20	+ +		+ +	+		+ +
Photoelectron-							
Spectroscopy	1–1000	+ +	+ +	+ +			+
Metallurgy	10^{a} 0.5^{b}	+ +	+	+	+ +		+
Biological/Medical:							
Microscopy	10	+ +	+	+ +	+		
Holography[c]		+ +	+ +			+	
Macromolecules,							
Crystallography	10	+ +	+	+	+ +	+	+
Medical,							
Radiography	1–10	+ +	+ +	+ +	+		+

[a] Surface studies.
[b] Volume studies.
[c] Coherence required.
[d] Line width in energy units.

4.2. Scientific

4.2.1. Atomic Physics and Accelerators

Some examples of possible applications of an x-ray laser to basic atomic physics include the following.

4.2.1a. PHOTOEXCITATION AND PHOTOIONIZATION. An x-ray laser can be used to pump other laser transitions at longer wavelength through processes discussed in Sections 3 of Chapters 3 and 5, respectively. The purpose would be to achieve other wavelengths and also to investigate the atomic kinetics of such pumping mechanisms. For example, it has been proposed[22] that a 2p electron in sodium be photoionized by the 206 Å and 209 Å selenium laser line outputs (Section 2 of Chapter 3) in a scheme to produce lasing on a 3s-2p self-terminating transition, as analyzed in Section 3.2 of Chapter 5.

4.2.1b. MULTIPLE PHOTOIONIZATION. Multi-charged ions are of great interest for accelerator sources of highly directional and high-power beams. Cool ions are advantageous over hot (plasma) ion sources. Multiple ionization by Auger cascade can occur when a K-shell electron is photoionized. In this process, for each K shell made vacant, two L holes are formed and from each of these two M holes result, etc. Double-Auger processes enhance this even further. Shake-off from the K-shell perturbation also releases electrons. For example, in mercury, the production of one L-vacancy has been shown to produce ions charged up to Hg^{17+}, which a mean charge of 9.8 + (mentioned in Ref. 8). Such cold ions are also of interest for additional lasing, due to the inverse dependence of gain on line width and hence on temperature for Doppler-broadened lines.

4.2.1c. PHOTO-POPULATION OF EXCITED STATES. Photoexcitation produces resonance-fluorescence effects in atoms and ions (discussed in the following paragraph). Saturation of an excited state allows a quantitative study of further excitation and ionization by either photons or electrons. This has important consequences. For example, collisional ionization from an excited state of an ion is the inverse process to dielectronic recombination. Hence, by detailed balancing, the dielectronic recombination rate may be deduced from the ionization rate.

4.2.1d. RESONANCE ABSORPTION AND FLUORESCENCE. Powerful methods of measuring ground-state population densities as well as excitation temperatures in atoms and ions exist, using absorption transitions matched by narrowband tunable lasers. A relatively straightforward measurement of the amount of radiation absorbed yields the density without absolute intensity measurements, when the absorption oscillator strengths are known. A well collimated laser beam is a particularly desirable source, permitting remote analysis[17] and not requiring large-solid-angle collection optics.

A further and more refined technique that can also lead to a measurement of the excitation temperature is to first saturate an absorption transition ending on an excited state. The next step is to measure the absolute fluorescence emission from that state, either in a different direction or to a different lower level. Assuming a statistical fractional population of the saturated upper state relative to the original ground state, the density in the latter can be determined. Alternatively, knowing the density from, e.g., an absorption measurement, the excitation temperature can be deduced.

Mostly to date such measurements have been carried out with neutral atoms, where the absorption transitions match the photon energy from "visible" tunable dye lasers. Such resonance absorption and saturation can also lead to greatly enhanced ionization through supra-elastic collisions, another application extending to short wavelengths. The technique is also important for plasma diagnostics, as described in Section 4.2.3.

4.2.1e. MINIATURE ELECTRON ACCELERATOR. Dubbed the "miniac", there is an interesting miniature accelerator concept proposed by Hofstadter[23] as a possible 1-cm version of the two-mile Stanford linear accelerator. The device would operate in crystals on discrete impulses from metastable atomic states generated by the incident coherent laser beams. The process of acceleration is the converse to an inelastic collision of an electron with an atom. An Auger process is one example of how excitation energy can be concentrated on a particle already in the atom. A coherent pump source such as a x-ray laser is required.

4.2.2. Electron Spectroscopy for Chemical Analysis (ESCA)

In this branch of surface science, one or more core electrons are photoionized by x-rays or released by Auger emission, and their kinetic energy distribution is analyzed[24,25]. This is used nondestructively to determine the elemental composition and chemical state of the surface of a specimen. It is important in the areas of metallurgy, catalysis and study of organic compounds. It is thus a highly promising application area for x-ray lasers[8].

The characteristics of an x-ray laser most needed for ESCA are the power output and the narrow linewidth. Also, an ~ 10 Å wavelength is desirable for core-level ionization. At this wavelength, a power output in the kW to MW range would be most adequate. A line width of < 1 eV is barely adequate. This corresponds to a line-width ratio of $\Delta v_{u\ell}/v_{u\ell} = 10^{-3}$ at $hv_{u\ell} = 1$ keV ($\lambda_{u\ell} = 12.4$ Å), i.e, about three times the Doppler broadening parameter adopted earlier. A width of 0.01–0.1 eV ($\Delta v_{u\ell}/v_{u\ell} = 10^{-4}$ to 10^{-5} at 1 keV) is actually desired, and would be possible with significant gain narrowing of the lasing line. Tunability over a few eV at an absorption edge could lead to interesting absorption spectroscopic investigations. Except for the link with linewidth, coherence and collimation are not essential requirements.

The potentially small beam diameter of the x-ray laser is both an advantage and a disadvantage. A number of microprobe exposures with a small x-ray laser beam followed by electron-microscopic analysis of the surface could yield useful information on the distribution and chemical state of surface elements. However, the high flux associated with the small beam could also destroy the sample, not an uncommon problem.

4.2.3. Diagnostics of High-Density Fusion Plasmas

In the field of compressed-pellet fusion, whether laser- or particle-driven, x-ray probing is essential to penetrate and diagnose the interior of the high-density plasma. Such plasmas are inertially compressed by a factor of ~ 5000 to supra-solid ion densities ($> 10^{27}$ cm^{-3}) and heated to $kT =$ tens of keV kinetic temperature. Probing with electromagnetic radiation requires a wavelength λ shorter than the critical value for absorption λ_p, determined by the plasma frequency[26]

$$v_p = \frac{1}{2\pi} \left[\frac{4\pi N_e e^2}{m_e} \right]^{1/2} \equiv \frac{c}{\lambda_p}, \tag{7}$$

where N_e is the electron density, and e and m_e are the charge and mass of the electron, respectively. Expressing wavelength in Angstrom units, this requirement for transmission becomes

$$\lambda \leq \lambda_p = \frac{3.4 \times 10^{14}}{(N_e)^{1/2}} \quad \text{Å}. \tag{8}$$

Hence, plasma electron densities of 10^{23}, 10^{25} and 10^{27} cm^{-3} require probing wavelengths shorter than 1000, 100 and 10 Å, respectively, i.e., in the xuv and soft x-ray spectral regions.

One important problem to be addressed is the stability of the interface between the fusion fuel and the pellet tamper, which ultimately affects the yield[8]. To address this question adequately, one needs x-ray images of the imploding plasma with μm and ps spatial and temporal resolution, respectively. These, along with collimation, are characteristics offered by x-ray lasers. Other vital information to be gained bears on energy absorption and compression history. Being an active and external probe, the x-ray laser would not be dependent on the state of the target.

One method of analyzing the compressed plasma is simply by absorption, i.e., the measured attentuation. In a pure fusion plasma, x-ray absorption would be due to free-free transitions for the electrons, rather than line absorption. This is sometimes referred to as inverse bremsstrahlung absorption.

For more quantitative determinations the plasma may be probed by (a) shadowgraphic, (b) interferometric, and (c) schlieren methods[27]. All three techniques depend on the bending of rays due to refraction in the plasma. In (a), the x-ray beam is simply bent by the plasma, and a pattern is formed if the gradient of the index of refraction varies over the plasma cross-section[28]. In (b), the plasma is placed between Bragg reflectors with parallel reflecting planes. The output beam is focused and then allowed to spread to an area detector. A knife edge at the focal point determines if and how the various beams deviated in the plasma medium. A careful analysis yields the electron density profile. This technique requires a small input aperture and therefore an intense and/or highly collimated source, i.e., an x-ray laser. In interferometric diagnostics (c), the plasma is placed in one arm of an interferometer, and the electron density is determined from the induced fringe shifts.

The resonance fluorescence technique described in Section 4.2.1d can be extended to plasma ions, using xuv and x-ray lasers. Such measurements then become very useful for fusion research as well as to further short wavelength plasma lasers and astrophysics. Of current fusion plasma interests, however, are practical and useful narrow-band tunable lasers operating at the absorption wavelengths of atomic hydrogen Lyman-α, β, γ, etc. ($n_i = 1$ to $n_f = 2, 3, 4$, etc.) in the 1000 Å region of the spectrum. For $n_f > 2$ fluorescence can be conveniently measured in the visible spectral region from n_f to $n = 2$ transitions. A wide range of hydrogen atom densities as well as temperatures can be measured by these techniques. Such emissions are currently produced by resonant four-wave sum-frequency mixing (FWSM) in a medium such as Hg-vapor, e.g., in which continuous tuning over a wide range of vuv

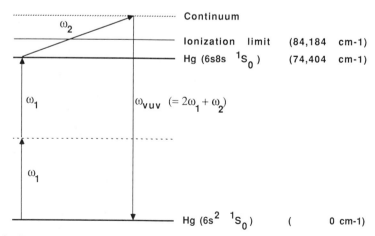

Fig. 6. Energy levels involved in a resonance fluorescence plasma diagnostic technique involving four-wave sum-frequency mixing. (From Ref. 29, reprinted courtesy Lawrence Berkeley Lab., U. Calif.)

frequencies is possible by changing the frequency of one of the long wavelength "visible" lasers used. This technique is illustrated[29] in Fig. 6. The process is coherent. When the medium (mercury vapor) is exposed to the laser electric fields of frequencies ω_1 and ω_2, some of the energy is converted to radiation at $2\omega_1 + \omega_2$. The process becomes resonant when $2\omega_1$ matches an absorption transition. Tuning is accomplished by ω_2 as indicated. The topic recently has been reviewed extensively[30] (see also Section 1 of Chapter 6).

4.2.4. Nuclear Decay Processes

Nuclear processes can be influenced by x-ray lasers through the surrounding electrons. The responsible processes are internal conversion and K-electron capture. The concentrated brightness of the x-ray laser is essential for these applications.

4.2.4a. INTERNAL CONVERSION.   In internal conversion an excited nucleus decays and expels an bound electron. This process can be suppressed by removal of some K-electrons by, e.g., photoionization. In a controlled experiment, this could serve as a diagnostic tool for studying the nuclear process in detail.

4.2.4b. K-CAPTURE.   During K-electron capture, an atomic electron is captured by a nucleus. It competes with positron beta-decay. An x-ray laser could conceivably inhibit K-capture and serve as a control on nuclear investigations. One observation would be a reduction of γ-ray emission.

4.2.5. Radiation Chemistry

The result of high energy radiation on matter eventually leads to chemical changes. The precise nature and evolution of such changes are of great interest[8]. The field involves in part photolysis using photons in the ultraviolet spectral region. It also involves radiolysis using typically 1 MeV gamma-rays. The region between could be well-served by a powerful x-ray laser source. The EXAFS (extended x-ray absorption fine structure) technique could be quite useful for quantitative determinations[31]. The goal is to study chemical entities in very short times (10^{-15}–10^{-12} sec), well suited to x-ray and/or gamma-ray laser characteristics. Such short bursts will be useful for the observation of short-lived excited states as well as for rapid initial reaction

rates in highly concentrated situations. It may also prove possible to disso-
ciate compounds with x-ray laser photons, and hence study the fundamental
chemistry of this process. Because most radiation chemistry is carried out
on liquid systems, wavelengths ≤ 1 Å are desired.

4.3. Technical

4.3.1. Grating and Grid Production with X-Ray Lasers

It may be possible to manufacture very high resolution gratings using x-
ray lasers. Gratings with sub-1000 Å spacings appear to be possible[32]. Such
gratings and grid structures can be used as diffractive and refractive optical
elements, as miniaturized circuit elements, and as research tools in basic
solid-state physics[33].

For generating x-ray transmission gratings of decreased groove spacing
for higher resolution, two methods have been suggested[21]. In the first, called
"spatial period division," harmonics of the diffraction pattern of a parent
transmission grating are used to produce a fringe pattern of improved resolution onto a photoresist. The x-ray laser beam does not have to be coherent,
but should be monochromatic and of high intensity. For a parent grating
of 2000-Å-wide spacing ($= 5000$ lines/μm), double and triple harmonics
would generate up to 15,000 lines/mm.

A second method of production is the holographic generation of reflection gratings. A split, monochromatic x-ray laser beam would be allowed to
recombine at the surface of a photoresist such as PMMA, giving an interference pattern. This technique requires a coherent beam, which is a definite
limitation for present x-ray lasers. The longitudinal coherence length limits
the line spacing to 1000 Å at present. However, 100-Å grid precision is anticipated[33]. Also, the quality of the grating is determined by the fringe contrast,
such that there is a further need to improve the longitudinal coherence by
as much as 100 times for a high quality reflection grating[21]. Nevertheless,
this remains a promising application.

4.3.2. Photolithography with X-Ray Lasers

High-resolution (< 1 μm) photolithography is one of the more promising
applications of x-ray lasers[8]. A series of comprehensive reviews on the status
of x-ray as well as electron- and ion-beam lithographies are published[34].
X-rays are preferred to ultraviolet radiation to reduce diffraction, scattering,

and reflection effects at sub-micron levels of resolution. The contact print[35] made from essentially point-projection photography must be recorded using a grainless medium. Negative photoresists such as PMMA (polymethylmethacrylate), in which polymerization is induced by the radiation, are chosen for high sensitivity (a comparison is included in Ref. 35). (See Section 6.4.1b of Chapter 2.) After development, the resist is removed in some areas, resulting in a relief image. The most desirable wavelength for achieving linewidths down to 1000 Å is in the 5–20 Å range, for minimum penetration and maximum contrast. Such small dimensions provide a high line-density for complex circuitry, increased reliability, and high frequency response.

Examples of particular micro-electronic devices appropriate for x-ray lithography fall into three categories:

(a) integrated circuits, diffraction gratings and magnetic bubbles on a large (> 5 mm^2) scale,
(b) transistors and surface acoustic wave devices (1 mm^2) on a moderate scale, and
(c) extremely small devices requiring 0.1 μm definition. This includes gratings for distributed-feedback solid state lasers, Josephson junction cryogenic devices and integrated optics.

Conventional x-ray sources require long exposures, i.e., tens of minutes to hours. Resolution is limited by the mask-to-substrate spacing required for good product yield, as well as the angular spread of the beam. Nevertheless, line-widths of the order of 200 Å have been reported[36]. The x-ray laser would provide the high intensity and collimation required. Also, the wavelength could be tailored to the absorption characteristics of the mask and resist materials. A highly coherent x-ray laser beam would also be advantageous for holographic or phase-contrast replications.

Plasma x-ray emission has been tested, short of lasing, for lithography. An example of a typical layout[37] is shown in Fig. 7. An exposure obtained with this setup is shown in Fig. 8.

One continuing problem is the threat of material damage due to a thermal spike from an intense short pulse of laser x-rays. Hence, the intensity must be controlled. Another problem is that the usually small beam must be expanded to cover a 1–10 cm diameter, while maintaining good collimation, for many lithographic applications. Finally there is the question of economics, which depends on the efficiency, size, cost, and repetition rate of the x-ray laser. With presently conceived devices, lithography would most likely be combined with other uses, much as synchrotrons and accelerators are used with multiple beam ports.

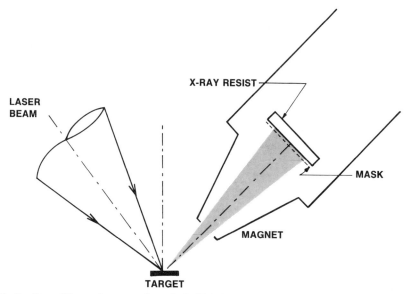

Fig. 7. X-ray lithography setup where the incident laser operated frequency doubled at 0.53 μm wavelength generates x-rays from a gold target. The focal spot size is 120 μm. (From Ref. 37.)

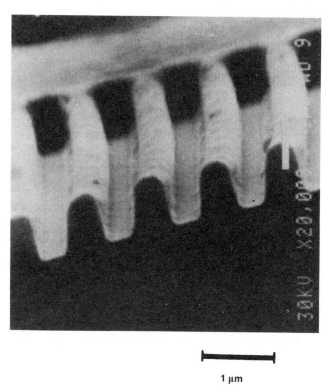

1 μm

Fig. 8. A 1 μm period line- and space-pattern generated in the setup shown in Fig. 7. (From Ref. 37.)

4.3.3. Photoelectron Spectroscopy

Photoelectron spectroscopy of a surface is a versatile uv and x-ray probe technique for studying intrinsic and absorbate surface states in solids[38]. Photons are used to release electrons from the surface, where their energy distribution is measured with an electron energy analyzer. In x-ray photoelectron spectroscopy, there is generally less surface sensitivity, and energy levels of core electrons of atoms are studied as well as valence electrons. Some questions to be addressed by this technique are

(a) What atomic species are present?
(b) What is their structural arrangement?
(c) How are the valence electrons distributed in space and in energy?
(d) What is the depth of adsorbed species in the specimen?

An x-ray laser provides the intensity needed in a collimated beam, as well as monochromaticity for controlling the depth of penetration. Soft x-rays are already useful as adjuncts to uv-probing.

4.3.4. Metallurgy

Nondestructive x-ray studies of metals fall into two categories: surface and volume.

4.3.4a. SURFACE STUDIES.   The study of metal surfaces does not require penetration, and x-ray lasers in the 10 Å range are useful. Some dynamical but irreversible processes occurring at and near the surface are[39]

(a) ion implantation and surface alloying,
(b) laser annealing,
(c) wear and friction,
(d) corrosion, and
(e) phase transformation.

The small size of the x-ray laser beam would be particularly useful in accurately determining surface dislocation cell structure and crack formation due to wear[8]. Such small-scale depth gauging requires operation at and across an absorption edge in the metal. Hence, the monochromatic and tunability characteristics of an x-ray laser are particularly attractive in discriminating against fluorescence.

Besides studying the above effects, the quality of the surface is of interest. An example is the evaluation of mirrors designed for focusing high power long-wavelength laser beams.

4.3.4b. VOLUMETRIC **S**TUDIES.   Volumetric studies require x-rays of <1 Å wavelength (>12 keV) for penetration of thin metallic layers. This is in the gamma-ray region (Section 3 of Chapter 6). For example, 0.7 Å (18 keV) radiation penetrates only 33 μm of iron. High power is also required to study interior flaws and their orientation.

4.4. Biological

The use of x-ray and gamma-ray lasers in the biological and closely related biomedical fields is one of the most challenging and promising possibilities.

4.4.1. X-Ray Microscopy

4.4.1a. BACKGROUND AND **N**EEDS.   It is very desirable to study biological cells and cell organelles with high spatial resolution and *in vivo*, i.e., in a natural and dynamic fluid state rather than dehydrated, stained and sliced[40,41]. This topic has been thoroughly reviewed in two conference proceedings[42,43]. Some vital questions[18] that can be addressed with an x-ray laser include:

(a) What are the locations of soluble enzymes in the cytoplasm, i.e., are they free-floating or associated with the cytoskeleton?
(b) Are protein-associated aggregated structures present within living cells?
(c) How does the cytoskeleton organize the structure of lipid membranes?

Time resolution on at least a sub-millisecond time scale allows visualization of cell morphology and rearrangement of the cytoskeleton during cell division[18]. Practical problems such as vibration and alignment drift will also be eliminated by the short exposures of an x-ray laser. With added coherence in an x-ray laser, three-dimensional holographic imaging of, for example, the structure of the cytoskeleton, becomes possible[18]. Holographic microscopy is described in more detail below.

With optical microscopes, the resolution achieved so far for such live studies is 500 Å with laser-produced plasma x-rays[4,44]. X-ray foci of 500–1000 Å have been achieved with Fresnel zone plates[45]. An improvement by ~ 100 times this can be achieved already with an electron microscope, but live or wet cells cannot be studied. One reason is that, in contrast to soft x-rays, electrons cannot penetrate the thickness of a living cell which is in the range of one to 10's of micrometers. Another problem with electron microscopy is that a living cell must remain in an atmospheric environment.

A 500-μm thick layer of air at atmospheric pressure has an x-ray transmittance of 70% at a wavelength of 45 Å. Electrons cannot penetrate such a layer. Hence, x-rays are needed to improve the resolution for live studies, with a reasonable goal[8,40] of 0.01 μm (or 100 Å). This approaches the virus, DNA and RNA scales. In this mode, x-ray microscopy will fill a much-needed gap between optical and electron techniques, and will supplement these classical methods.

There are three major areas of technical challenge in x-ray microscopy:

(a) a sufficiently intense x-ray source,
(b) the high-resolution imaging optics required,
(c) the contrast mechanism, i.e., the interaction between the x-rays and the substance irradiated.

For (a), the x-ray laser with its intensity and collimation offers enormous opportunities. In fact, this may indeed prove to be the most exciting application. Repetitive operation is highly desirable. The imaging required (b) is discussed in Section 6.1 of Chapter 2 and shows great promise of rapid improvement, particularly using layered synthetic microstructure (LSM) reflecting optics described in Section 2.3.2 of Chapter 1.

4.4.1b. OPERATION NEAR THE SPECTRAL "WATER WINDOW". Concerning (c) above, it is of special interest that proteins, lipids, etc., in the live biological sample are more absorbing than the water present for wavelengths shorter than 43.6 Å (the wavelength below which carbon absorbs strongly, i.e., the K-absorption edge). This high contrast continues to wavelengths as short as 23.30 Å, which is the corresponding oxygen K-absorption edge in the water. This wavelength range, i.e., 23.3–43.6 Å (and perhaps slightly longer) has become known as the "water window" for high-contrast live biological x-ray spectroscopy. This is illustrated[45] in Fig. 9. Absorption coefficients are also tabulated in Ref. 46. The long wavelength limit of this window can be extended somewhat by intentional doping of the sample with high-Z substances for good contrast on a localized basis[47], even though this distorts the specimen composition. For example, it has been suggested[21] that, with the attachment of Fe to anti-bodies which then become attached to specific cellular structures ("tagging"), the wavelength can be extended to about 60 Å.

This high-contrast spectral "window" requirement has served to set one of the most important goals in terms of short wavelength for x-ray lasers. Gain in this wavelength region is being rapidly approached (see Tables in Sections 2 of Chapters 3 and 4). No doubt there will continue to be many

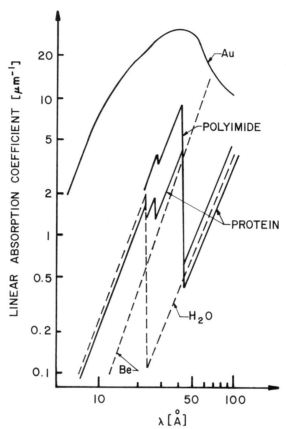

Fig. 9. Linear absorption coefficients for cellular protein and water showing the "water window" effect. Also included is a plastic film (polymide) and a light (Be) and heavy (Au) metal, for comparison. (From Ref. 45.)

further successes near this spectral region in the near future, followed by improved beam quality and fascinating new biological discoveries.

4.4.1c. RADIATION DAMAGE TO THE SAMPLE.   An important question in high-resolution and high-brightness live x-ray microscopy concerns radiation damage[48]. With intermediate resolution at least, it has already been demonstrated that is possible to observe living cells[40,45,49]. However, with an x-ray laser source, there is little doubt that the specimen will receive the lethal dosage of x-rays (~ 1000 J/cm^2, $\propto \lambda_{u\ell}^{-4}$) required to destroy cellular structures[35] at the highest resolution (100 Å). However, the fine structure of the cells is expected to remain intact. Some near-term information on the threshold dosage can be obtained from contact microscopy with existing x-ray lasers[21]. Such a threshold is of great importance for future x-ray holography of living cells, described below. In such contact microscopy, the sample is irradiated by a x-ray laser without any intermediate imaging optics[35]. The developed photoresist is then scanned by an electron microscope. Resolutions of approximately 100 Å have been achieved using PMMA as a resist[50]. There is no requirement here for coherence.

4.4.2. X-Ray Holography

If there is a single application that most justifies the development of coherent x-ray lasers it is biological holography. A true x-ray hologram could, in principle, provide time-resolved three-dimensional images of the live cell. A comprehensive study of the problems and possibilities peculiar to biomicroholography with x-ray lasers is available[46]. X-ray holograms taken to date with essentially point sources have suffered from a lack of adequate brightness and coherence, i.e., the number of photons per spatial mode is small, so very long exposure times are required.

Probably the least difficult and hence first approach for useful x-ray laser holography will be Gabor holography, where the object is placed between the source and the detector, with x-rays passing the object serving as the reference beam. This is described in more detail in Ref. 21. The design of a high brightness, single mode, spatially coherent x-ray oscillator/amplifier combination suitable for this purpose is described in Ref. 6. Prior to the development of such a laser, the 200 Å selenium laser described in Section 2 of Chapter 3 is currently used in a simple Gabor in-line geometry to generrate holograms of 8-μm carbon fibers and a 10-μm gold wire array[5].

In the meantime, the present technology in biological analysis has been advanced to a resolution of 400 Å, using an undulator on an electron storage ring as a source of 24.7 Å radiation, within the "water window" region[51]. Holograms of zymogen granules from rat pancretic cells were obtained on high-resolution photoresists exposed for one hour. This dramatizes the huge advantage that is promised by a bright, nanosecond duration x-ray laser in this field, for in-vivo cellular studies.

4.4.3. Macromolecules

To understand better the bio-molecular basis of life, it is important to examine individual macromolecules in living tissue, which is exceedingly difficult[52]. Moreover, the atoms involved have extremely small refractive indices at x-ray wavelengths, and hence give poor contrast with conventional x-ray sources. With a coherent x-ray laser, the wavefronts from each scattering atom are related, such that variations of the scattered wavefront due to small changes in the refractive index could be observed using phase-contrast microscopy[53]. X-ray holograms, discussed above, would allow three-dimensional observation of such atoms.

4.4.4. Crystallography

A potentially very significant application of x-ray lasers is the study of micron-sized crystals[8,52]. This takes advantage of the high power and small beam size available. Such a technique would have a particular impact on biomolecular crystallography. It would immensely simplify the task of growing crystals free of irregularities if it could be done on such a small scale, compared to mm-sized crystals analyzed with conventional x-ray sources.

A coherent x-ray laser could also be used to solve the "phase problem" in x-ray scattering in crystals[53]. Solution of this problem is essential to determining the electron distribution in a crystal. In ordinary x-ray sources, the wavefront is not well defined in space or time. With such an x-ray laser, the phases of the scattered waves could be determined with relative ease.

4.5. Medical Radiology

Radiological medicine is an area of applications that comes immediately to mind in evaluating x-ray lasers for practical application in clinical diagnostics and general physiological studies[8]. This is in part because of the enormous expansion in use of long-wavelength lasers in both clinical and research environments in recent years. So far, practical use has been limited by the lack of adequate x-ray sources. High brightness, monochromaticity, collimation, coherence and an exposure time varying from 0.1 sec to as short as a few milliseconds for some angiographic applications are required. X-ray and gamma-ray lasers promise to fill this need. They are also elemental to further improvements in sensitivity and resolution in the relatively new and exciting field of computerized tomography (CT-scanning) with x-ray fluoroscopy.

4.5.1. Differential Absorption

Differential absorption can be used to ascertain quantitatively the distribution of heavy elements in various organs in-vivo and with improved contrasts[54]. Such a technique has general applications in situations in which a pathological condition has led to a change in elemental composition. Some familiar examples are iodine in the thyroid gland and osteoporosis in bones. For example, a difference of absorption of less than 1% for two wavelengths lying close but on opposite sides of an absorption edge of the element can be measured[55]. Iodine content can be determined with an accuracy of at least $\pm 20\%$[56]. For osteoporotic patients, a quantitative measurement by such a method of the distribution of calcium or bone mineral content in the bone could be extremely useful. The physiology of muscle contraction and control can also benefit from x-ray diffraction at high intensity[57].

This technique requires intense and penetrating radiation of high monochromaticity such as expected from a laser. Also, for *in-vivo* use, this method of analysis requires penetration through centimeters of tissue and hence photon energies in excess of 10 keV, i.e., in the gamma-ray laser regime (see Section 3 of Chapter 6).

4.5.2. Holographic Imaging

Holography imaging in the true sense as understood and expected from lasers could have tremendous clinical potential for diagnosis of abnormal anatomical situations caused by congenital abnormalities or disease. Examples include the location, analysis and extent and geometry of tumor masses with respect to normal structures, as well as the dynamic changes associated with the functions of the various organ systems.

The holographic procedure for recording three-dimensional images of incoherently illuminated objects originated[58-60] in the late 1960s. The holographic stereogram consisted of a single multiplexed plate made from a series of ordinary photographic transparencies. A large patient dosage was required. Later, tomographic techniques were incorporated to reduce the required exposure. This technique has been used to produce three-dimensional representations of lumbar vertebrae[61]. This technique again requires gamma-ray laser photon energies for penetration. However, softer x-ray lasers could also provide rapid elemental analysis of cells by microphotometry and fluorescence.

4.5.3. Microprobing

There are at least two medical areas which make use of an extremely small diameter collimated x-ray laser beam[8]. The first involves studies of effects of small beams on a living cell. Single-cell absorption measurements can be carried out. The distribution of angular scattering can also lead to cell identification. The second area involves studies which use the beam as a tool to analyze cell function, by irradiating various subcellular structures. For this, beam diameters on the 0.25 μm (2500 Å) scale are required.

4.6. Summary

In summary, the x-ray laser has a brilliance of $\sim 10^8$ times that of the x-ray sources now known (see Fig. 4). This is most dramatic for applications requiring high spectral brilliance, and hence warrants the further development of such a new and powerful tool. The most exciting applications are as yet probably unimagined and hence unpredictable. It is safe to predict that they will be as revolutionary in the x-ray region as they have been with visible and infrared lasers. Six orders of magnitude in brilliance accompanied by coherence could not possibly go unappreciated and unused for any significant time. The best is yet to come for the development and application of x-ray lasers.

5.   PROGNOSIS

X-ray laser research and development has advanced to the point where it is today through a series of spurts of interest and progress, with a periodicity of $\sim 3-5$ years beginning around 1970. Some of the periodically renewed interests have arisen with the availability of new and more powerful drivers. Some have followed a breakthrough in thinking and small-scale experiments. Mostly, however, they have resulted from changes in administrative policies at various levels.

Now that there exist lasers in the xuv and soft x-ray spectral regions of useful output power, the next logical step is to "institutionalize" this area of research, rather than continuing as an adjunct to laser fusion and pulsed power programs. Fully dedicated, world-wide cooperative facilities are needed. Such laboratories must attract the best talent available. As such, they must be equipped with theoretical/numerical capabilities along with complete diagnostics, in order to develop a clear understanding of the physics and engineering involved during further development.

Likely areas of fruitful pursuit include, in an unprioritized order:

- shorter wavelength,
- increased output energy and power,

- longer pulse, cavity capability,
- improved beam quality,
- reduced size,
- enhanced efficiency, and
- improved convenience and accessibility.

Clearly the promise is exciting for x-ray laser research and development in the coming years. The truly revolutionary capabilities and applications are limited only by our imaginations and available resources.

REFERENCES

1. M. H. Key, *J. de Physique* **49**, C1-135 (1988).
2. D. Matthews, et al., *J. Opt. Soc. Am. B* **4**, 575 (1987).
3. S. Suckewer, C. H. Skinner, H. Milchberg, C. Keane and D. Voorhees, *Phys. Rev. Letters* **55**, 1753 (1985).
4. M. H. Key, *Nature* **316**, 314 (1985).
5. J. E. Trebes, et al., *Science* **238**, 517 (1987).
6. M. D. Rosen, J. E. Trebes and D. L. Matthews, *Comments in Plasma Physics and Controlled Fusion* **10**, 245 (1987).
7. D. J. Nagel, *Annals New York Acad. Sci.* **342**, 235 (1980).
8. S. Jorna, et al., "X-ray Laser Applications Study," Report No. PD-LJ-77-159 (Physical Dynamics, Inc., LaJolla, California 1977).
9. V. V. Mikhailin, S. P. Chernov and A. V. Shepelev, *Sov. J. Quantum Electron.* **8**, 998 (1978).
10. J. A. R. Samson, "Techniques of Vacuum Ultraviolet Spectroscopy" (Pied Publications, Lincoln, Nebraska, 1967).
11. M. J. Weber, in "Energy and Technology Review—Synchrotron Radiation," p. 1 (Lawrence Livermore National Laboratory, Livermore, California, Nov.-Dec. 1987).
12. "Handbook on Synchrotron Radiation," E.-E. Koch, ed (North Holland Publ. Co., Amsterdam, 1983).
13. M. R. Howells, "Optical Components and Systems for Synchrotron Radiation: An Introduction," SPIE Proc. No. 315: Reflecting Optics for Synchrotron Radiation, p. 13 (1981).
14. R. C. Elton, T. N. Lee and P. G. Burkhalter, *Nuclear Instruments and Methods in Physics Research* **B9**, 753 (1985).
15. R. C. Elton, T. N. Lee and W. A. Molander, *Phys. Rev. A* **33**, 2817 (1986).
16. J. P. Apruzese, et al., *SPIE Proc.* **875**, 2 (1988).
17. R. A. Andrews, "X-Ray Lasers—Current Thinking," Memorandum Report No. 2677 (U.S. Naval Research Laboratory, Washington, DC, 1973); summarized in Laser Focus **9**, 41 (Nov. 1973).
18. Proceedings of the First Symposium on the Applications of Laboratory X-Ray Lasers, N. M. Ceglio, ed., Report CONF-850293 (Lawrence Livermore National Laboratory, Livermore, California, 1985), (abstracts and vugraphs; video tape also available).
19. R. W. Waynant and R. C. Elton, *Proc. IEEE* **64**, 1059 (1976).
20. D. J. Nagel, "Potential Characteristics and Applications of X-ray Lasers", in Advances in X-ray Spectroscopy, C. Bonnelle and C. Mande, eds., Chapter 20 (Pergamon Press, New York, 1982).
21. J. Trebes, *J. de Physique* (Colloque) **C6**, 309 (1986).

22. W. T. Silfvast, O. R. Wood, II, and D. Y. Al-Salameh, in AIP Conference Proceedings No. 147, D. T. Attwood and J. Boker, eds., p. 134 (American Institute of Physics, New York, 1986).
23. R. Hofstadter, "The Atomic Accelerator," Report No. 560 (Stanford Univ. High Energy Physics Lab. Stanford, California, 1968).
24. K. Siegbahn, et al., "ESCA-Atomic, Molecular and Solid State Structure Studied by Means of Electron Spectroscopy" (Almquist and Wiksells, Uppsala, 1967).
25. T. A. Carlson, "Photoelectron and Auger Spectroscopy" (Plenum Press, New York, 1975).
26. S. Glasstone and R. H. Lovberg, "Controlled Thermonuclear Reactions," p. 171 (D. Van Nostrand Publ., New York, 1960).
27. S. L. Leonard, in "Plasma Diagnostic Techniques," R. H. Huddlestone and S. L. Leonard, eds., Chapt. 2 (Academic Press, New York, 1965).
28. S. Nakai, et al., in "Advances in Inertial Confinement Systems," C. Yamanaka, ed., p. 90 (Osaka Univ. Press, 1980).
29. G. C. Stutzin, et al., *Rev. Scientific Instruments* **59**, 1363 (1988).
30. C. R. Vidal, in "Topics in Applied Physics, Vol. 59: Tunable Lasers," L. F. Mallenauer and J. C. White, eds., p. 57 (Springer-Verlag, Berlin, 1987).
31. J. Bordas, in "Uses of Synchrotron Radiation in Biology," H. B. Stuhrmann, ed., Chapter 6 (Academic Press, New York, 1982).
32. G. C. Bjorklund, S. E. Harris and J. F. Young, *Appl. Phys. Letters* **25**, 451 (1974).
33. P. L. Csonka and R. Tatchyn, AIP Conference Proceedings No. 75, p. 304, D. T. Attwood and B. L. Henke, eds. (American Institute of Physics, New York, 1981).
34. *SPIE Proceedings* **773**, "Electron-Beam, X-Ray, and Ion-Beam Lithographies VI, P. D. Blais, ed., 1986 and VII, A.W. Yanof, ed., 1989.
35. W. Gudat, in "Uses of Synchrotron Radiation in Biology," H. B. Stuhrmann, ed., Chapter 2 (Academic Press, New York, 1982).
36. D. C. Flanders, *Vac. Sci. Technol.* **16**, 1615 (1979).
37. L. Mochizuki, R. Kodama, C. Yamanaka, and H. Aritome, "Laser Plasma X-Ray Source for Lithography," SPIE Proc. No. 773: Electron-Beam, X-Ray, and Ion-Beam Lithographies VI, p. 246 (1987).
38. D. E. Eastman and M. I. Nathan, "Photoelectron Spectroscopy," in *Physics Today* **28**, 44 (April 1975).
39. T. Sparks, in Proceedings of the First Symposium on the Applications of Laboratory X-Ray Lasers, N. M. Ceglio, ed., Report CONF-850293, p. 174 (Lawrence Livermore National Laboratory, Livermore, California, 1985).
40. G. Schmahl, D. Rudolph and B. Niemann, AIP Conference Proceedings No. 75, D. T. Attwood and B. L. Henke, eds., p. 225 (American Institute of Physics, New York, 1981); also "Uses of Synchrotron Radiation in Biology," H. B. Stuhrmann, ed., Chapter 3 (Academic Press, New York, 1982).
41. V. E. Cosslett and W. C. Nixon, "X-ray Microscopy" (Cambridge U. Press, 1960)
42. D. F. Parsons, ed., *Annals New York Acad. Sci.* **342**, 1–401 (1980).
43. J. Kirz and D. Sayre, in "Uses of Synchrotron Radiation," H. Winick and S. Domiack, eds., p. 277 (Plenum Press, New York, 1980).
44. R. W. Eason, et al., *Optica Acta* **33**, 501 (1986).
45. G. Schmahl, D. Rudolph, B. Niemann and G. Christ, in "Ultrasoft X-Ray Microscopy: Its Application to Biological and Physical Sciences," D. F. Parsons, ed., *New York Academy of Sciences* **342**, 368 (1980); and "X-ray Microscopy," p. 63 (Springer-Verlag, 1984).
46. J. C. Solem and G. C. Baldwin, *Science* **218**, 229 (1982).
47. R. Feder and D. Sayre, *Annals New York Acad. Sci.* **342**, 213, (1980).
48. A. Halpern, in "Uses of Synchrotron Radiation in Biology," H. B. Stuhrmann, ed., Chapter 10 (Academic Press, New York, 1982).

49. D. Sayre, J. Kirz, R. Feder, D. M. Kim and E. Spiller, *Ultramicroscopy* **2**, 337 (1977).

50. J. W. McGowan, B. Borwein, J. A. Medeiros, T. Beveridge, J. D. Brown, E. Spiller, R. Feder, J. Topalian and W. Gudat, *J. Cell Biology* **80**, 732 (March 1979).

51. M. Howells, C. Jacobsen, J. Kirz, R. Feder, K. McQuaid and S. Rothman, *Science* **238**, 514 (1987).

52. H. D. Bartunik, R. Fourme and J. C. Phillips, in "Uses of Synchrotron Radiation in Biology," H. B. Stuhrmann, ed., Chapter 7 (Academic Press, New York, 1982).

53. A. Ferguson, *New Scientist* **68**, 207 (Oct. 23, 1975).

54. B. J. Panessa-Warren, *Annals New York Acad. Sci.* **342**, 350, (1980).

55. N. A. Baily and R. L. Crepeau, *Radiology* **115**, 439 (1975).

56. B. Jacobson, *Amer. J. of Roentgenol. Rad. Ther. and Nucl. Med.* **91**, 202 (1964); also P. Edholm and B. Jacobson, *Acta Radiologica* **52**, 337 (1959).

57. R. S. Goody and K. C. Holmes, in "Uses of Synchrotron Radiation in Biology," H. B. Stuhrmann, ed., Chapter 8 (Academic Press, New York, 1982).

58. J. T. McCrickerd and N. George, *Appl. Phys. Letters* **12**, 10 (1968).

59. J. D. Redman, *J. Sci. Instr.* **1**, 821 (1968).

60. E. Shuttleworth, A. Wilson, J. D. Redman and W. P. Walton, *Brit. J. Radiol.* **42**, 152 (1969).

61. N. A. Baily, E. C. Lasser and R. L. Crepeau, *Invest. Radiol.* **6**, 221 (1971).

List of Symbols

The following symbols are introduced in the text at the page numbers indicated.

Arabic Symbols

a	Cross sectional area of the laser medium, 53
a_L	Area of irradiated lasing plasma in photon coupling, 137
a_0	Bohr radius, 101
A	Einstein coefficient for spontaneous emission, 23
A_0	Radiative decay rate following dielectronic capture, 170
$A_{u\ell}$	Transition probability for the laser transition, 23
$A_{\ell f}$	Transition probability for the depletion transition, 36
A_{uo}	Transition probability for the pumping transition, 53
A_{uf}	Transition probability for the upper laser state, 173
$B_{u\ell}$	Einstein coefficient for induced emission, 23
B_w	Wiggler field in FEL design, 212
b	Confocal parameter, 206
c	Velocity of light in vacuum, 23
$C_{\ell u}$	Excitation rate coefficient from level "ℓ" to "u", 56
C_{ud}	Excitation rate coefficient from level "u" to a d-level, 108
d	depth of medium, 38
d_B	lattice spacing for Bragg diffraction, 86
D_{coh}	Extent of region of transverse coherence, 244
D_m	Illuminated breadth for mirror as focusing element, 66
D_u	Total decay rate from level "u", 27
e	Charge of electron, 39
E_e	Electron energy in a beam, 232
$E_{pk}(\text{i-i})$	Peak energy in ion-ion charge transfer, 178
$\mathscr{E}$	Optical field, 201
f	(subscript) Final state after lasing and depletion, 19
f_{dw}	Debye-Waller factor for gamma-ray lasers, 218
$f_{f\ell}$	Absorption osc. strength for the depletion transition, 39
$f_{\ell u}$	Absorption osc. strength for the laser transition, 23
$f^{\#}$	f-number of a Fresnel zone plate, 80

274

N_i	Density of ions, 3
N^i	Density of atoms in ion stage "i", 60
	Population density of bound electrons:
N_f	in final laser level after depletion, 19
N_ℓ	in lower-laser level, 18
N_o	in immediate state from which pumped, 19
N_s	in source state (preceding "o"), 19
N_u	in upper-laser level, 18
N_y	Density of neutral-atoms "Y" in charge transfer, 176
$\mathscr{N}_c$	Total number of reflections (or passes) in a cavity, 30
$\mathscr{N}_e$	Number of equivalent electrons, 183
$\mathscr{N}_w$	Number of wiggler periods in FEL design, 211
$\mathscr{N}_z$	Total number of zones in a Fresnel zone plate, 80
$\mathscr{N}_v$	Number of photons 23
o	(subscript) Immediate state from which pumping occurs, 19
P_{ou}	Pumping rate from level o to level u, 27
P_{ce}	Rate for electron-collisional excitation pumping, 101
P_{ct}	Rate for charge transfer pumping, 176
P_{dr}	Rate for dielectronic-recombination pumping, 170
P_{cr}	Rate for collisional-recombination pumping, 148
P_{pe}	Rate for photoexcitation pumping, 127
P_{pi}	Rate for photoionization pumping, 163
P.I.	Population inversion, 155
$\mathscr{P}$	Polarization, 201
q	Harmonic order, 207
r	Radius of laser medium, 16
r_a	Characteristic absorption dimension in radiation coupling, 139
r_n	Radius of the nth ring in a Fresnal zone plate, 80
r_0	Classical electron radius, 23
r_e	Radius of electron orbit in synchrotron, 249
R_{cr}	Collisional-recombination rate coefficient, 60
R_{ct}	Charge transfer rate coefficient, 60
R_{dr}	Dielectronic-recombination rate coefficient, 60
R_∞	Rydberg constant for hydrogen, in wave numbers, 188
R_m	Radius of mirror in focusing optics, 64
R_{rr}	Radiative-recombination rate coefficient, 60
R_1, R_2	Reflectances for two cavity mirrors, 30
$R_=$	Mean reflectance for $R_1 = R_2$, 76
Ry	Rydberg constant for hydrogen, in energy units, 37
s	(subscript) Source level preceding level "o", 19
S_{ou}	Ionization rate coefficient creating upper-laser state, 183
$\mathscr{S}_d$	Shape function for a Doppler-broadened line, 25
$\mathscr{S}_s$	Shape function for a Stark-broadened line, 24
$\mathscr{S}_x$	Shape function for an arbitrary line shape, 24
t_{coh}	Duration of temporal coherence, 245
t_g	Duration of gain, 75
T_e	Electron temperature in plasma, 2
T'_e	Reduced electron temperature in plasma, 56
T_i	Ion temperature in plasma, 2
T_B	Brightness temperature for blackbody radiator, 43
u	(subscript) Upper laser state, 18
v_{th}	Thermal velocity in plasma, 40

v_x	Relative atom-ion velocity in charge transfer, 176
$v_x(\text{i-i})$	Relative velocity in ion-ion charge transfer, 176
V	Volume, 137
w	Width of line image, 66
W	Power radiated, 52
$W^{(n)}$	Power in nth harmonic, 206
W_p	Pumping power for photoexcitation, 137
x	Separation of plasmas in radiation coupling, 138
$\langle x \rangle$	Characteristic (mean) distance in radiation coupling, 139
X^{i+}	i-times ionized atom of element "X", 100
X_s^{i+}	General source ion of charge $i+$, 161
X_o^{i+}	General pumped ion of charge $i+$, 161
X_n^{i+}	General ion of charge $i+$ in intermediate quantum state n, 162
X_u^{i+}	General upper-laser-state ion of charge $i+$, 162
Y	General perturbing neutral atom in charge transfer, 174
z	Distance measured along laser axis, 19
z_m	Separation of mirrors in a cavity, 75
Z	Nuclear charge, 42
Z_p	Nuclear charge for pumping ion, 42
Z_{abs}	Nuclear charge for absorbing ion, 45
Z_x	Nuclear charge for general ion X in charge transfer, 175

Greek Symbols

α	Fine structure constant, 148
α_{ic}	Internal conversion coefficient, 218
β_{rel}	Relativistic factor in gamma-ray laser design, 232
γ	Velocity gradient in a flowing plasma, 39
γ_w	Reduced energy for wiggler design in free-electron lasers, 211
γ_{rel}	Relativisitic Lorentz factor in gamma-ray laser design, 229
Γ_{ic}	Internal conversion decay rate for gamma-ray lasers, 218
Γ_o	Autoionization rate following dielectronic capture, 170
δ_r	Outermost Fresnel zone width, 80
Δ	Reduced depth of medium, 58
ΔE_{ud}	Energy difference: level "u" to a higher d-level, 108
$\Delta E_{u\ell}$	Energy difference between laser levels, 36
ΔE_{so}	Energy difference between dielectronic-capture levels, 171
$\Delta\lambda$	Line width in wavelength units, 23
$\Delta\lambda_d$	Line width for a Doppler-broadened line, 25
$\Delta\lambda_s$	Line width for a Stark-broadened line, 25
$\Delta\lambda_x$	Line width for an arbitrary shape, 24
Δn_a	Change in principal quantum no. for absorbing transitions, 132
Δn_p	Change in principal quantum no. for pumping transitions, 139
Δr	Change in radial direction, 50
$\Delta\tau_c$	Variation in opacity for streaming plasma, 40
Δv	Line width in frequency units, 23
Δv	Variation in streaming velocity in flowing plasma, 40
Δv_s	Relative streaming velocity in counter-flowing plasmas, 141
ε	Ratio of densities between the "f" and "u" levels, 56
ε_0	Permittivity of free space, 201
ζ	Net charge on ion, 36

η	Geometric factor for coupled plasmas, 138
θ_B	Incident angle for Bragg diffraction, 86
θ_m	Off-axis tilt angle in mirror focusing of driver laser, 66
θ_r	Refraction angle, 50
κ_c	Absorption coefficient at line center, 38
λ	Wavelength, 5
λ_{abs}	Wavelength of the absorbing transition, 47
λ_c	Wavelength at the center (peak) of a spectral line, 25
λ_{crit}	Critical wavelength in synchrotron spectrum, 248
λ_d	Wavelength in Doppler broadening estimates, 25
λ_{depl}	Wavelength for the depletion transition, 47
λ_i	Wavelength for incident field in harmonic generation, 206
$\lambda_{\ell f}$	Wavelength of the lower-final transition, 39
λ_L	Laser wavelength in free-electron laser and channeling, 211
λ_p	Critical plasma wavelength for absorption, 63
$\lambda_{u\ell}$	Wavelength of laser transition, 19
λ_{uo}	Wavelength of pumping transition, 128
λ_{uv}	Wavelength of incident ultraviolet in harmonic generation, 204
λ_w	Wiggler spatial periodicity in free-electron laser design, 212
μ	Atomic mass number, 25
μ_a	Dipole moment of a single atom, 15
ν	Frequency of radiation, 6
ν_{uo}	Frequency of the pumped transition, 52
ν_p	Critical plasma frequency for absorption, 257
ζ	Correction for incomplete nuclear screening, 188
ρ	Occupation index in blackbody formula, 128
ρ_r	Radiation density for induced emission, 23
σ_{abs}	Line absorption cross section, 22
σ_{ce}	Electron-collisional excitation cross section, 56
$\sigma_{ce}(i\text{-}i)$	Excitation cross section in ion-ion collisions, 185
σ_{cs}	Compton scattering cross section for gamma-ray lasers, 220
σ_{ct}	Charge transfer cross section, 176
σ_{dr}	Dielectronic (recombination) capture cross section, 170
σ_{iict}	Ion-ion charge transfer cross section, 178
σ_{ou}	Excitation cross section from level "o" to "u", 101
σ_{pe}	Photoexcitation cross section, 127
σ_{pi}	Photoionization cross section, 28
σ_{stim}	Stimulated emission cross section, 21
σ_{24}	Cross section for excitation from level 2 to level 4, 42
τ_c	Optical depth at line center, 38
ϕ	Beam divergence angle, 242
Φ	Focusing and dispersion factor in harmonic conversion, 206
χ	Ionization potential, 148
χ_y	Ionization potential of atom Y in charge transfer, 175
$\chi^{(n)}$	Electric dipole susceptibility of order n, 201
Ψ_{cav}	Cavity enhancement factor, 76
ω_h	Frequency of harmonic, 202
ω_i	Frequency of incident field for harmonic generation, 202
ω_n	Frequency of nth specific incident field for harmonics, 202
Ω	Solid angle, 138

Index